THE
POWER OF
WHEN

Learn the best time
to do **EVERYTHING**

MICHAEL BREUS, PhD

Vermilion
LONDON

1 3 5 7 9 10 8 6 4 2

Vermilion, an imprint of Ebury Publishing,
20 Vauxhall Bridge Road,
London SW1V 2SA

Vermilion is part of the Penguin Random House group of companies
whose addresses can be found at global.penguinrandomhouse.com

Copyright © Michael Breus 2016

First published in the United Kingdom by Vermilion in 2016
First published in the USA by Little, Brown 2016

www.penguin.co.uk

A CIP catalogue record for this book is available from the British Library

ISBN 9781785040450

Printed and bound in Great Britain by Clays Ltd, St Ives PLC

Penguin Random House is committed to a sustainable future for
our business, our readers and our planet. This book is made from
Forest Stewardship Council® certified paper.

As my daughter told me, this book "should be dedicated
to your awesome children and wife." I could not have
put it any better myself.
The book is dedicated to my Wolfpack: Lauren, Cooper, Carson,
and my four-legged kids: Monty, Sparky, and Sugar Bear.
And a special dedication to all my patients throughout my
sixteen years of practice. I love learning from you all,
every time we meet.

Contents

Contents

PART THREE
THE POWER OF WHEN FOR LIFE

Foreword

Dr. Breus has been a friend and colleague of mine since my show began. His tireless enthusiasm for learning, and for educating the public and being out in front of cutting-edge information about sleep and sleep disorders, has made him one of the core experts in many of my ventures.

I became interested in the healing power of circadian rhythms when I was in a meeting room with Dr. Breus. We were discussing the future of sleep medicine and why sleep deprivation was the most underappreciated health and wellness problem in America. I wanted to know what the next big thing in sleep would be.

Dr. Breus explained that the circadian system, also known as your biological clock, affects every area of functioning in the body, controlling everything from the multiplication of cancer cells to the integrity of the immune system. It became clear to me that there was a wealth of research on the topic but little exposure to it among the general public. I knew that people needed to be educated about it in a meaningful way, and so I encouraged Dr. Breus to write *The Power of When*.

The more you understand circadian dyssynchrony—a concept presented in the book in a highly accessible way—the more you will improve your life. For example, the gut has a circadian pacemaker of its own. When your gut is not on its biological clock schedule, hormone disruption causes increased levels of inflammation, inefficient metabolism, even decreased effectiveness of many prescription therapies.

The quiz at the beginning of the book helps you figure out which one of four distinct chronorhythm groups you fall into. Then you get a basic understanding of what a typical day for a Lion, Bear, Wolf, or Dolphin looks like and "when" the best time to do many basic tasks is.

As many of you know, I am a big proponent of regular bowel movements, and I speak about them frequently on my show. One of my favorite chapters in the book explains "when" to have one; it does not get much easier than that! Another chapter that I paid specific attention to was "when" to take medication effectively so that it can improve your life — overnight. And what about physical activity? Dr. Breus has dedicated an entire section of this book to identifying times "when" you will get the most benefit and pleasure from being active.

Circadian science will advance medical testing. Testing can now become more precise through the time-stamping of specimen collection and the comparison of results to time-based norms. Clinicians can get more accurate results. What if your blood was drawn to look at your thyroid levels in the morning versus the evening: Could the results differ to the point of diagnosis? It appears so.

Based on a simple understanding of your biology and scheduling, you will learn "when" you can get the most out of yourself and your significant relationships, in areas like sex, love, planning an event, and talking to your kids. "When" you see improvement in these areas, you can enhance your health and your life in ways you could never have imagined.

Of course, we cannot forget about work, which occupies so much of our time. Knowing and understanding "when" you function best, and "when" others function best, allows you to put your best ideas forward, be most creative, and be open to instruction. This book will allow you to learn "when" you can truly be your best on many levels.

I took the quiz and discovered I am a Lion. I identified well with this chronotype's characteristics, and realized that I had unknowingly created a schedule that worked for me, with many areas right on target. But I decided to change the timing of napping in my life to see if I could make this particular area more efficient. I was amazed at the impact this one simple change had on my health. And so it gives me great pleasure to write this foreword to tell all of you how this book can help you, your family, your career, and your health.

—Mehmet C. Oz, MD

THE
POWER OF
WHEN

Timing Is Everything

Do you want a simple, straightforward life hack that requires little effort and gets you closer to happiness and success? Of course you do! This might sound like a promise waiting to be broken. It's not.

You've probably already seen a lot of tricks and tips about the "what" and "how" of success.

How to lose weight.

How to please a sexual partner.

What to say to your boss to get a pay increase.

How to raise your kids.

What to eat.

How to work out.

What to think.

How to dream.

"What" and "how" are excellent and necessary questions. But **there is another crucial question that *must* be addressed in order to make fast, dramatic, lasting improvements in the quality of your life across the board.**

That question is "when."

"When" is the ultimate life hack.

It's the foundation of success, the key that unlocks a faster, smarter, better, and stronger you.

Knowing "when" enables you to perform "what" and "how" to your maximum potential. If you didn't change a thing about *what* you do and *how* you do it, and only made micro-adjustments to *when* you do it, you'd be healthier, happier, and more productive, starting...right now.

"When" really is that simple, and that powerful.

Just by making small tweaks to your schedule—such as when to have the first cup of coffee, when to answer emails, when to nap—you'll nudge the rhythm of your day back in sync with the rhythm of your biology, and then everything will start to feel easier and flow naturally.

What do I mean by the "rhythm of your biology"?

Contrary to what you might have heard, there *is* a perfect time to do just about anything. Good timing isn't something you choose, guess, or have to figure out. It's already happening inside you, in your DNA, from the minute you wake up to the minute you fall asleep, and every minute in between. An inner clock embedded inside your brain has been ticking away, keeping perfect time, since you were three months old.

This precisely engineered timekeeper is called your circadian pacemaker, or biological clock. Specifically, it's a group of nerves called the suprachiasmatic nucleus (SCN), in the hypothalamus, right above the pituitary gland.

In the morning, sunlight comes into your eyeballs, travels along the optic nerve, and activates the SCN to begin each day's circadian (Latin for "around a day") rhythm. The SCN is the master clock that controls dozens of other clocks throughout your body. Over the course of the day, your core temperature, blood pressure, cognition, hormonal flow, alertness, energy, digestion, hunger, metabolism, creativity, sociability, and athleticism, and your ability to heal, memorize, and sleep, among many other functions, fluctuate according to and are governed by the commands of your inner clocks. Everything you can do or want to do is controlled by physiological rhythms, whether you realize it or not.

For fifty thousand years, our ancestors organized their daily schedules around their inner clocks. They ate, hunted, gathered, socialized, rose, rested, procreated, and healed on perfect bio-time, or biological time. I'm not saying life was fantastic in prehistoric, biblical, or medieval times, but as a species, we thrived by rising with the sun, spending most

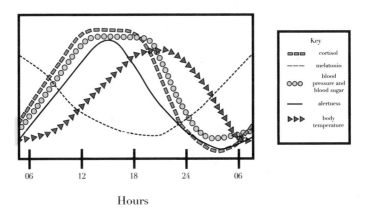

Hours

This is an example of several circadian rhythms going on inside you RIGHT NOW!

of the day outdoors, and sleeping in total darkness. We created civilization and societies and made incredible advances that, ironically and effectively, turned our finely tuned and evolved inner clocks against us.

The most disruptive event in the history of bio-time occurred on December 31, 1879. At his research lab in Menlo Park, New Jersey, Thomas Edison introduced the long-lasting incandescent lightbulb to the world. He famously said, "We will make electricity so cheap that only the rich will burn candles." Within a decade, night, for all intents and purposes, became optional. We no longer rose at dawn and slept in total darkness. We once worked from dawn to dusk and ate our last meal in twilight. Working hours and dinnertime shifted later and later. We spent more time indoors exposed to artificial light and less time outdoors under the sun.

In an 1889 interview with *Scientific American*, Edison said, "I hardly ever sleep more than four hours per day, and I could keep this up for a year."[1] In 1914, on the thirty-fifth anniversary of incandescent light, Edison used the occasion to identify sleep as a "bad habit." He proposed that all Americans sleep far fewer hours per day, and predicted a future of sleeplessness. "Everything which decreases the sum total of man's sleep increases the sum total of man's capabilities," he said. "There really is no reason why men should go to bed at all, and the man of the future will spend far less time in bed."[2]

The second major disruption of your biological time was transportation advances. Cars and planes allowed people to travel great distances rapidly. It takes a day for the body to adjust to a one-hour time zone difference, and, on horseback or in a coach, it'd take about that long to go that far. Starting in the mid-twentieth century, in the blink of an eye, evolutionarily speaking, we could travel multiple time zones in a few hours, leaving bio-time lagging behind.

Computer technology brought us to where we are now, in a 24/7 smartphone culture of perpetual dusk where we work, play, and eat around the clock.

It took only 125 years to undo 50,000 years of perfect bio-timekeeping. Saying that our physiology hasn't evolved as quickly as our technology is the understatement of the millennium. As a result, our "when" is way, way off.

Being out of sync with bio-time is devastating to one's physical, mental, and emotional well-being. The phenomenon is called chrono-misalignment ("chrono" means time). In the last fifteen years, scientists have been connecting the so-called diseases of civilization (mood disorders, heart disease, diabetes, cancer, and obesity) with chrono-misalignment. Symptoms include insomnia and sleep deprivation, which lead to depression, anxiety, and accidents, to say nothing of what feeling overwhelmed and exhausted does to relationships, careers, and health. Unless you turn off every screen and light at 6:00 p.m., you are likely to deal with chrono-misalignment in one way or another, whether it's morning fogginess, extra weight, feeling stressed out, or not performing to your potential. (It's unrealistic to power down at dusk, of course. But you can turn off screens a bit earlier than usual, and dim the lights as the night goes on.)

A sparrow doesn't rush to work by 9:00 a.m. on a coffee buzz while fighting traffic. A salmon doesn't attend a midnight concert. A deer doesn't binge-watch *House of Cards* all weekend. Imagine a house cat napping, playing, and cleaning on a societal schedule. It would never happen. Animals heed their inner clocks. Humans, with our big, superior brains, willfully ignore ours, cramming our circadian rhythms into a "social rhythm," often in direct opposition to what our bodies are supposed to be doing at that time.

HOW I DISCOVERED THE POWER OF WHEN

I became board-certified in clinical sleep medicine fifteen years ago, around the same time that chronobiology (the study of circadian rhythms) became *the* hot topic in my field. The study of circadian rhythms in humans was virtually nonexistent before the 1970s and is still unknown to the general public. Why? For starters, most primary care doctors haven't heard of chronobiology, either. It isn't taught in medical school, except when it comes to a few rare sleep disorders. There's no official drug to prescribe for being out of sync with your inner clock (unless you count caffeine, the most abused substance on the planet!), but there are many drugs and nutraceuticals—foods that have a medicinal value— that have a detrimental effect on your bio-time. (For a list of such drugs, go to page 160.)

When a number of my patients weren't responding to standard insomnia therapy, I became interested in, and then fascinated with, chronobiology. I had to branch out and find new ways to help them, and I started using chronotherapy techniques—exposure to light boxes at certain times of the day, replacing lightbulbs in the bedroom with "sleep-friendly models" (check out Lighting Science, www.lighting.science) and recommending the "sleep hormone" melatonin at specific times within the circadian cycle—with some measure of success.

But then I wondered if my patients would get even better results if they adjusted their daily schedules to inch closer to their natural bio-time. I asked them to make minor changes to when they ate and exercised, spent time with friends and watched TV, and were exposed to artificial light. When they did, we started to see remarkable improvement, not only in their sleep but in their general health, mood, memory, concentration, fitness, and weight. Good timing is so powerful, I realized, that it could change *anyone's* life in almost every way.

I was hooked. I read everything I could find in medical journals about the profound benefits of being in sync with your bio-time. As I mentioned, the field has exploded as a research topic, so I have been very busy keeping up. Here's just a small sample of the top studies to make circadian breakthroughs in the last few years:

- **Treating a disease such as cancer on bio-time can save your life.** In 2009, researchers from the University of North Carolina School of Medicine experimented with mice to determine if timing of medication affected the speed of DNA repair to damaged cells. They took extracts of the mouse brains at various times and found that when medication was taken at night, **DNA repaired itself seven times faster,** in correspondence with the circadian rising and falling levels of a certain enzyme. The researchers theorized that, to minimize side effects and maximize effectiveness, chemo drugs should be given to patients when their cells are better able to repair themselves.

- **Thinking on bio-time can make you smarter and more creative.** In 2011, a team of psychologists from Michigan State University and Albion College asked their study subjects to solve problems, some analytical and some that required insight, at different times throughout the day. **The subjects solved creative problems better during their non-optimal times, when they were tired and groggy. They solved analytical problems at their optimal times, when they were wide-awake and alert.** The researchers concluded that creative and analytical thinking operates on bio-time. If you set out to solve a certain type of problem, you'll do better at certain times.

- **Eating on bio-time can help you manage weight.** In a 2013 study of 420 overweight or obese men and women, researchers at the University of Murcia, Spain, put the subjects on a diet of 1,400 calories per day for twenty weeks. Half of the subjects were "early eaters," having their biggest meal of the day before 3:00 p.m. The other half, the "late eaters," had their biggest meal after 3:00 p.m. The two groups ate the same quantities of the same food, exercised at similar intensity and frequency, slept the same number of hours, and had comparable appetite hormones and gene function. Which group lost more weight? **The early eaters lost twenty-two pounds, on average; the late eaters lost, on average, seventeen pounds, a 25 percent difference.** The late eaters were more likely to skip breakfast.

- **Living on bio-time can make you happier.** In 2015, researchers at Copenhagen University Hospital, in Denmark, treated seventy-five patients with major depression through the use of either daily chronotherapy (bright light exposure and a consistent wake time) or exercise. **Sixty-two percent of the chronotherapy patients went into remission in six months.** Only 38 percent of the exercisers did.

- **Running on bio-time can make you faster.** In 2015, a team from the University of Birmingham, England, set out to find a connection between athletes' performance and whether they felt alert and active in the morning (morningness) or alert and active in the evening (eveningness). There is one, indeed. **The number of hours between a runner's wake time and race time had a huge impact on performance. If the late risers ran in the evening, they were much faster than if they ran in the morning, for example. The differences in speed were significant, measuring up to 26 percent.**

In the pages ahead, you'll read more about these and many other studies. They offer proof of the importance of keeping good bio-time, and demonstrate the dangers of ignoring it. The scientific fact is, if you are time-wise, your life will tick along like clockwork.

If you are out of sync with your inner timing, you're working against your own biology. When has that ever been a good idea?

I'm not an Edison hater. I'm not about to say you have to throw away your iPhone or go live in a cave. If not for science and technology, we wouldn't have the proof of just how profound bio-timing is to health and productivity. We can use the research and technology to help us keep near-perfect bio-time and still stay on a social schedule. That's the beauty of it: You don't have to overhaul your life to tap into the power of when. You only have to shift a few things around, set up some alarms on your phone, download my free app, and watch your life change for the better.

WATCHWORDS

Bio-time: Your biological clock or schedule; the ebb and flow of hormones and enzymes, and the changes in circulatory activity over the course of the twenty-four-hour day. Synonymous with "circadian rhythm."

Chronobiology: The study of circadian rhythm and its effect on human health and wellness.

Chrono-misalignment: The negative impact on your health, focus, and energy when your social schedule is out of sync with your biological schedule.

Chronotherapy: Using tools like light and hormonal supplements to improve health and quality of life for patients with insomnia and mood disorders.

Chronotype: A classification of the general timing of your biological clock.

Chronorhythm: A schedule of the optimal physiological time to do just about every daily activity within the context of our busy, modern lives. This is a daily rhythm for success.

Circadian rhythm: Your biological clock or schedule; the ebb and flow of hormones and enzymes, and the changes in circulatory activity over the course of the twenty-four-hour day.

Social jet lag: The foggy feeling you get when your social schedule is out of sync with your biological schedule.

Social schedule/social rhythm: When you do things—rise, eat, exercise, work, socialize—throughout the day.

Time-wise tips and tricks: Strategies that help you sync your social schedule and your biological schedule.

CHRONOTYPES

CHAPTER 1

What's Your Chronotype?

Every person has a master biological clock ticking away inside his or her brain, and dozens of smaller biological clocks throughout his or her body.

But not every person's biological clocks keep the same time. Your friend's inner clocks might run at a different pace than yours, or your partner's, or your kids'. You know this already; you've observed that some people wake early, or don't feel hungry when you do, or are full of energy when you are winding down. Different people fall into different classifications, called "chronotypes," based on general morningness and eveningness preferences.

According to conventional wisdom and historical definition, there are three chronotypes:

1. **Larks,** the early risers
2. **Hummingbirds,** neither early nor late risers
3. **Owls,** the late risers

Psychologists and sleep doctors have long used a standard Morningness-Eveningness Questionnaire (MEQ) to determine an individual's chronotype. Having worked with patients and studied in the field for over fifteen years, I'd always been bothered by the three categories and how they were determined. By only assessing an individual's sleep/wake/activity preferences, the MEQ didn't match the patient population in my clinical practice *at all*.

The established chronotype assessment didn't include both measures of the two-step system for sleep. Along with wake preference, there is "sleep drive" — your need for sleep. Some people have higher sleep drives than others, just as some have stronger sex drives than others.

Your sleep drive is genetic, and it determines how much sleep you need and your depth of sleep.

Those with **low sleep drive** don't need a lot of sleep, so the night seems very long to them. Low sleep drive people are easily woken up by sound and light disturbance, and they wake up feeling less than refreshed.

Those with **high sleep drive** need more hours of sleep, so the night feels too short for them. High sleep drive people sleep deeply, but they wake up feeling less than refreshed no matter how much sleep they get.

Those with **medium sleep drive** sleep somewhat deeply and are satisfied and refreshed by seven hours of continuous rest.

The MEQ wasn't designed to take into account the individual's personality. But personality turns out to be incredibly important for figuring out chronotype. For example, morning types tend to be more health-conscious. Evening types tend to be impulsive. Neither type tends to be easygoing. This has been confirmed in dozens of studies. In a comprehensive evaluation of chronotype, personality is too big and relevant to ignore.

My second issue was that classic definitions didn't match up with my patient population. The three established types excluded 10 percent or more of the general population: insomniacs. Although bad sleepers can be found among early, late, and normal risers, true insomniacs — those who chronically struggle to fall and/or stay asleep and usually get less than six hours per night — I believe are a distinct chronotype, with a wake/sleep preference, sleep drive, and personality profile that are distinct from those of the classic three categories.

I decided to redefine the groups and write a questionnaire of my own that took all the important factors into account. I also renamed the chronotypes. Humans are mammals, not birds, and we share behaviors with other mammals. My chronotype names reflect that. I looked for mammals that accurately represented the four categories as I see them, and found exactly what I was looking for:

1. **Dolphins.** Real dolphins sleep with only half of their brain at a time (which is why they're called unihemispheric sleepers). The other half is awake and alert, concentrating on swimming and looking for predators. This name fits insomniacs well: intelligent, neurotic light sleepers with a low sleep drive.

2. **Lions.** Real lions are morning hunters at the top of the food chain. This name fits morning-oriented driven optimists with a medium sleep drive.

3. **Bears.** Real bears are go-with-the-flow ramblers, good sleepers, and anytime hunters. This name fits fun-loving, outgoing people who prefer a solar-based schedule and have a high sleep drive.

4. **Wolves.** Real wolves are nocturnal hunters. This name fits night-oriented creative extroverts with a medium sleep drive.

If you don't recognize yourself in the short descriptions above, perhaps you recognize one of your parents. Remember, your chronotype is genetic—determined specifically by the PER3 gene. If you have a long PER3 gene, you need at least seven hours of deep sleep to function, and tend to be an early riser. If you have a short PER3, you can get by on light or little sleep, and you tend to be a late riser. It's likely that at least one of your parents had the same chronotype as you.

Why so many types? Why is there variation at all? Since the dawn of man, a range of chronotypes has been necessary for species survival. Each chronotype had its purpose and contributed to the larger group's security. Bio-time had to be diverse for the larger group to stay safe over the long night. Although we don't stand watch over the cave opening anymore, our genetic structure hasn't changed all that much since prehistoric times, and neither have the following ratios:

- **Dolphins account for 10 percent of the population.** Light sleepers, they rouse at the smallest noise to wake and warn the group of danger.
- **Lions account for 15 to 20 percent.** They rise early, taking the morning shift of guarding the group and watching out for roving predators.

- **Bears account for 50 percent.** Their cycles match the rise and fall of the sun; they hunt and gather in daylight.
- **Wolves account for 15 to 20 percent.** They take the late shift to guard the group, drifting off when the most extreme Lions start to stir.

Obviously, these four types don't run on the same bio-time. For example, a Lion's metabolism does not match up with a Wolf's, so it doesn't make sense for a Lion to eat on a Wolf's schedule. For optimal health and performance, each type has its own chronorhythm, or daily schedule for success. In upcoming chapters, I'll provide a detailed chronorhythm for each type.

Generally speaking, Dolphins, Lions, and Wolves are naturally out of sync with social norms, and their chronorhythms reflect that. Bears' bio-time is the closest to established norms. They're the largest group, which explains how social norms were created in the first place. However, just because the norms exist does not mean they're helping Bears reach their creative, professional, and personal goals. Bears will notice that their chronorhythms call for adjustments to the way they live now as well.

You probably have a pretty good idea which chronotype you are already. Time to confirm your suspicions by taking the Bio-Time Quiz (BTQ) on page 17. It incorporates every important factor, including sleep/wake/activity preference and personality, as well as taking into account behavioral tip-offs and observations I've made among my patients. The BTQ has been tested and retested on several populations—my patients, the general public, selected friends and colleagues—and is the most accurate tool for assessing chronotype I was able to imagine and create.

It is comprised of two parts. Part One is a series of ten statements to be marked True or False. Part Two is a series of twenty multiple-choice questions. There are no right or wrong answers. Try to be as honest and objective as possible when answering. (Relax, Wolves. You won't be graded on this. And, no, Lions, there's no perfect score.)

If you want to do this on your phone or computer, go to www.the powerofwhen.com.

BIO-TIME QUIZ

Part One

For the following ten statements, please circle "T" for True or "F" for False.

1. **The slightest sound or light can keep me awake or wake me up.**
   T or F
2. **Food is not a great passion for me.**
   T or F
3. **I usually wake up before my alarm rings.**
   T or F
4. **I can't sleep well on planes, even with an eye mask and earplugs.**
   T or F
5. **I'm often irritable due to fatigue.**
   T or F
6. **I worry inordinately about small details.**
   T or F
7. **I have been diagnosed by a doctor or self-diagnosed as an insomniac.**
   T or F
8. **In school, I was anxious about my grades.**
   T or F
9. **I lose sleep ruminating about what happened in the past and what might happen in the future.**
   T or F
10. **I'm a perfectionist.**
    T or F

If you marked "T" for True for **seven or more** of the above ten statements, **you are a Dolphin** and can skip ahead to page 22.

Otherwise, continue on to . . .

Part Two

After each of the multiple-choice options, you'll find a number in parenthesis. Keep a tally of these numbers to get your final score.

1. **If you had nothing to do the next day and gave yourself permission to sleep in as long as you like, when would you wake up?**
   a. Before 6:30 a.m. (1)
   b. Between 6:30 a.m. and 8:45 a.m. (2)
   c. After 8:45 a.m. (3)

2. **When you have to get out of bed by a certain time, do you use an alarm clock?**
   a. No need. You wake up on your own at just the right time. (1)
   b. Yes to the alarm, with no snooze or one snooze. (2)
   c. Yes to the alarm, with a backup alarm, and multiple snoozes. (3)

3. **When do you wake up on the weekends?**
   a. The same time as your workweek schedule. (1)
   b. Within forty-five to ninety minutes of your workweek schedule. (2)
   c. Ninety minutes or more past your workweek schedule. (3)

4. **How do you experience jet lag?**
   a. You struggle with it, no matter what. (1)
   b. You adjust within forty-eight hours. (2)
   c. You adjust quickly, especially when traveling west. (3)

5. **What's your favorite meal? (Think time of day more than the menu.)**
   a. Breakfast (1)
   b. Lunch. (2)
   c. Dinner. (3)

6. **If you were to flash back to high school and take the SAT again, when would you prefer to *start* the test for maximum focus and concentration (not just to get it over with)?**
   a. Early morning. (1)

   b. Early afternoon. (2)

   c. Midafternoon. (3)

7. **If you could choose any time of day to do an intense workout, when would you do it?**

   a. Before 8:00 a.m. (1)

   b. Between 8:00 a.m. and 4:00 p.m. (2)

   c. After 4:00 p.m. (3)

8. **When are you most alert?**

   a. One to two hours post wake-up. (1)

   b. Two to four hours post wake-up. (2)

   c. Four to six hours post wake-up. (3)

9. **If you could choose your own five-hour workday, which block of consecutive hours would you choose?**

   a. 4:00 a.m. to 9:00 a.m. (1)

   b. 9:00 a.m. to 2:00 p.m. (2)

   c. 4:00 p.m. to 9:00 p.m. (3)

10. **Do you consider yourself...**

   a. Left-brained—that is, a strategic and analytical thinker (1)

   b. A balanced thinker (2)

   c. Right-brained—that is, a creative and insightful thinker (3)

11. **Do you nap?**

   a. Never. (1)

   b. Sometimes on the weekend. (2)

   c. If you took a nap, you'd be up all night. (3)

12. **If you had to do two hours of hard physical labor, like moving furniture or chopping wood, when would you choose to do it for maximum efficiency and safety (not just to get it over with)?**

   a. 8:00 a.m. to 10:00 a.m. (1)

   b. 11:00 a.m. to 1:00 p.m. (2)

   c. 6:00 p.m. to 8:00 p.m. (3)

13. **Regarding your overall health, which statement sounds like you?**

   a. "I make healthy choices almost all of the time." (1)

    b. "I make healthy choices sometimes." (2)

    c. "I struggle to make healthy choices." (3)

14. **What's your comfort level with taking risks?**

    a. Low. (1)

    b. Medium. (2)

    c. High. (3)

15. **Do you consider yourself:**

    a. Future-oriented with big plans and clear goals. (1)

    b. Informed by the past, hopeful about the future, and aspiring to live in the moment. (2)

    c. Present-oriented. It's all about what feels good now. (3)

16. **How would you characterize yourself as a student?**

    a. Stellar. (1)

    b. Solid. (2)

    c. Slacker. (3)

17. **When you first wake up in the morning, are you . . .**

    a. Bright-eyed. (1)

    b. Dazed but not confused. (2)

    c. Groggy, eyelids made of cement. (3)

18. **How would you describe your appetite within half an hour of waking?**

    a. Very hungry. (1)

    b. Hungry. (2)

    c. Not at all hungry. (3)

19. **How often do you suffer from insomnia symptoms?**

    a. Rarely, only when adjusting to a new time zone. (1)

    b. Occasionally, when going through a rough time or are stressed out. (2)

    c. Chronically. It comes in waves. (3)

20. **How would you describe your overall life satisfaction?**

    a. High. (0)

    b. Good. (2)

    c. Low. (4)

SCORING

19 to 32: **Lion**
33 to 47: **Bear**
48 to 61: **Wolf**

DO HYBRIDS EXIST?

Sometimes, people take the quiz, read the profiles, and are still uncertain which type they fall into. Within each major type (Lion, Bear, and Wolf), there is a range. But even if some Bears wake up earlier than other Bears, that does not make them Lions.

If you are straddling Lion/Bear or Wolf/Bear, you are probably a Bear, like the majority of the population.

To further hone your assessment, try a two-question mini-quiz devised by Brazilian researchers[1] that is as accurate as any other standard measurement.

1. Rate your energy level on a scale of 1 (very low) to 5 (very high) in the morning.
2. Rate your energy level on a scale of 1 to 5 in the evening.

Subtract the second score from the first score. For example, if you rated the morning energy as very high (5) and the evening energy as very low (1), your total score is 4. If you rated your morning energy as very low (1) and your evening energy as very high (5), your total score is -4.

SCORING

4, 3 or 2: **Lion**
1, 0, -1: **Bear**
-4, -3, -2: **Wolf**

The greatest chronotype confusion is usually over **the insomnia**

issue. Are all insomniacs Dolphins? Not necessarily. Each major type has its bad sleepers and shares some personality traits with Dolphins. Some extreme Lions wake up at 3:00 a.m., can't fall back to sleep, and have been told by their doctor that they have "sleep offset insomnia." Lions, like Dolphins, are conscientious, goal-oriented, and risk-averse. Extreme Wolves stare at the ceiling every night until 3:00 a.m., and their doctors have called this "sleep-onset insomnia." Wolves, like Dolphins, are introverted, creative, and anxious. And some Bears are often irritable and frequently fatigued. There are some similarities but also important differences.

If you suspect you might be a Dolphin, even if you got six or fewer "True" answers on Part One of the BTQ, take a follow-up mini-quiz (if not, skip ahead to page 26):

Are You a Lion or a Dolphin?

Answer T or F to the following statements.

1. I'm not very hungry when I wake up.
   T or F
2. My sleep is fitful and shallow.
   T or F
3. I have no interest in being the boss.
   T or F

If you answered T to at least two of the above three statements, you are a Dolphin.

Are You a Bear or a Dolphin?

Answer T or F to the following statements.

1. You don't really care about food.
   T or F
2. You'd be thrilled to get six hours of sleep a night.
   T or F

3. You aren't a team player.

T or F

If you answered T to at least two of the three statements, you are a Dolphin.

Are You a Wolf or a Dolphin?

Answer T or F to the following statements.

1. You are usually the last to leave a party.

    T or F
2. You're spontaneous and make snap decisions about major purchases and vacation plans.

    T or F
3. You hit snooze at least twice in the morning.

    T or F

If you answered F to at least two of the above three statements, you are a Dolphin.

BEAR FACTS

Bears might think that they have it easy. Since their schedules are solar-synced, that must mean they don't have many changes to make to their daily schedules to be on perfect bio-time, right?

Wrong.

Let me ask you, Bears: Are you operating at peak potential on your current schedule? Are you bursting with energy and getting high-quality sleep? Are you carrying extra weight around the middle? Having as much energetic sex as you'd like? Killing it at work? Enjoying excellent communication in relationships? Avoiding colds and the flu? Able to focus and concentrate as well as you'd like?

Just because your sleep/wake cycle is on bio-time does not mean that the dozens of other clocks in your body are in sync. Established social schedules—work time, dinnertime, sex time—do *not* automatically match up with a Bear's chronobiology.

In fact, Bears, you have a lot of adjustments to make to your daily schedules. But the effort will pay off. Wouldn't it be nice to feel sharp at work all day, to avoid late-night fridge raids, to wake up refreshed, not to be dependent on coffee to wake up and alcohol to fall asleep, to feel alive and healthy every day of the week? Of course!

If you make my recommended changes to your daily schedule, you take control of your destiny and become the person you were born to be.

LION ENVY

A close friend of mine once took this quiz and said, "I'm a Bear."

Knowing him to be a weekend napper, a foodie, and a personable guy with many friends (and a few extra pounds), I wasn't surprised to hear it. But he was.

"I don't want to be a Bear!" he said. "I want to be a Lion!"

My friend does have some Lion traits. He started up his own business and sees himself as a real go-getter with intense ambition and drive. His professional goals are his prime motivator, and, having heard me talk about chronotypes, he sized himself up, based on this one aspect of his personality, as a Lion.

If you see yourself at the top of the food chain and are disappointed not to actually be a Lion, or if you aspire to be an early riser with energy to spare and a strategic mind, know that your Lion Envy is misplaced. Each chronotype has advantages and disadvantages across the board, from careers to relationships to physical health. What might seem like a bonus could be a detriment. Lions do tend to rise through the ranks and become bosses, but they tend not to be creative and extroverted. Lions might get more done before breakfast than most of us do all day, but they have a hard time in the social arena due to their very early bedtime and fatigue.

The grass is always greener on the other side of the chronotype. Instead of wishing you were another type, develop self-awareness and understanding of your own bio-time patterns.

One question I'm asked often: **Can you change your chronotype?**

Chronotype is genetic. It's in your DNA. You can't change your chronotype, just as you can't change your height or eye color. But you can make subtle shifts of one or two hours within the natural range of your chronotype's bio-timing. If you were born a Lion, you will never naturally be able to stay up as late as a born Wolf. But by adjusting the timing of meals, exercise, caffeine, and exposure to artificial and natural light, each chronotype stands to make huge improvements in health, energy, and productivity. Bears will have more energy and will be able to lose weight and get as much sleep as they need; Lions will stay awake later and enjoy a richer social life. Wolves will fall asleep earlier and become more productive in the mornings and early afternoons, and Dolphins will improve the quality of their sleep, quiet their anxiety, and be more productive earlier in the day.

So the answer to the question "Can you change your chronotype?" is no.

However, your chronotype can, and will, change on its own, depending on how old you are. (Go to "Chrono-Longevity," on page 327, for a detailed explanation of why and when this happens.) Toddlers tend to be Lions. Teenagers tend to be Wolves. Adults tend to be Bears. Seniors tend to be Lions and Dolphins. You won't be the same chronotype for your *entire* life. But between the ages of twenty-one (when teen Wolves transition into, for the most part, Bear adults) and sixty-five (when Bear adults transition into Dolphins and Lions), the forty-plus years that are considered the prime of your life, your chronotype will be the same—and you should get on the right bio-time for whatever chronotype you are right *now*.

CHRONOTYPE PROFILES

When researchers conduct scientific studies, their conclusions are based on percentages, not absolutes. For example, when researchers in the department of biology at the University of Education in Heidelberg, Germany, surveyed 564 students in 2014 to study chronotype and personality traits, they found that the majority of evening types rated high in "impulsivity," and that the majority of morning types rated high in "activity." This does not mean that *every* Wolf is a low-energy sensation-seeker, and that *every* Lion is a cautious go-getter. Of course not. But most are.

A 2012 study from the National Institute of Health and Welfare in Helsinki, Finland, found that morning types eat more fish, vegetables, and whole grains, while evening types consume more soda and chocolate. But you might be a Diet Coke–loving Lion or a sushi-addicted Bear. A particular trait or preference in your chronotype profile might not describe or fit you exactly, but, as a whole, you will recognize yourself. I've given this quiz to hundreds of people, and, according to my subjects' estimations, the profiles are 80 to 100 percent accurate. I hear the phrases "spot-on" and "scarily true" more often than not.

Everyone is studying this topic. The research I've mined to create these profiles is drawn from all over the world. The connection between chronotype and mood comes from the University of Poland.[2] The link between eveningness and impulsivity comes from Yonsei University College of Medicine, in Seoul, South Korea.[3] The study on perceived health and chronotype was a joint effort by researchers from London, Budapest, and Cork, Ireland.[4] I've included data from Chicago and Bangkok,[5] Madrid[6] and Padua.[7] My point in noting the far-reaching scope of the research isn't to show off but to prove that chronotypes are the same all over the globe. Lions are Lions from New York to Hong Kong. Bears are Bears from Scottsdale to Calcutta. Wherever you go on Earth, you will find your kind. Chronotypes, unite!

DOLPHIN

- **Four Key Personality Traits:** Cautiousness, introversion, neuroticism, intelligence
- **Four Key Behaviors:** Avoiding risky situations, striving for perfection, obsessive-compulsive tendencies, fixating on details
- **Sleep/Alertness Pattern:** Dolphins usually wake up feeling unrefreshed and are tired until late in the evening, when they suddenly hit their stride. Most alert: late at night. Most productive: in spurts throughout the day. Naps: They try to nap to catch up on sleep but can't quite make it happen.

Dolphins in nature are unihemispheric sleepers. One half of the brain shuts down while the other half stays alert to prevent drowning and being eaten by predators. Their feeding pattern is flexible, and their metabolism is rapid.

The human equivalent is a light sleeper, easily roused by slight sounds and disturbances. Dolphins have a low sleep drive, and struggle with waking up multiple times over a long night. Being a light sleeper and having a low sleep drive can cause anxiety-related insomnia. As they lie awake at night, Dolphins can ruminate about mistakes they've made and things they've said, how they would redo their mistakes and misstatements if they could, what they can expect to deal with in the future, and how they'll accomplish anything on so little rest. When Dolphins do sleep, they skim the surface and often claim that they don't know whether or not they actually slept at all.

Out of bed, Dolphins are generally anxious as well. They tend to have Type A personalities—nervous, irritable, and worried—and are highly intelligent. Their attention to detail and perfectionism are ideally suited for precision work—copyediting, coding, engineering, chemistry, composing or performing music. Their obsessiveness and perfectionism do not equip them for teamwork. They are happiest (or least irritable) when left alone to work by themselves and do what they need to do. They often get mired in the details—to the point where they become paralyzed and get little done.

Dolphins tend to be less guarded in their relationships, and they tend to demonstrate exceptional emotional attention (listening and being present), repair (the ability to fix problems), and clarity (understanding what's really going on). Confrontation-averse, Dolphins don't enjoy arguing, but at times they are so sleep-deprived that they may have marital conflict anyway. Lingering emotional tension is even more troubling than arguing and is to be avoided at all costs. In my clinical observation, Dolphins might not be the easiest people for Bears and Lions to live with, but they make attentive, caring, dedicated partners. I've noticed that Dolphins and Wolves seem to make a good combination.

Regarding their health, Dolphins are usually eat-to-live types with a naturally fast metabolism. Some are obsessive exercisers, but most Dolphins don't really care about fitness and don't need to exercise in order to lose weight, since their body mass index (BMI) tends to be low to average. Although Dolphins will try any over-the-counter or prescription medication for their insomnia or hypochondria, they can be obsessively careful about what they eat, drink, and purchase. They rate themselves low in overall life satisfaction. At the same time, being vocal about their dissatisfaction makes them feel better. Check out your perfect day on page 34.

LION

- **Four Key Personality Traits:** Conscientiousness, stability, practicality, optimism
- **Four Key Behaviors:** Overachieving, prioritizing health and fitness, seeking positive interactions, strategizing
- **Sleep/Alertness Pattern:** Lions wake up bright-eyed at dawn or earlier, start to feel tired in the late afternoon, and fall asleep easily. Most alert: noon. Most productive: morning. Naps: Lions hardly ever nap. They'd rather be doing something useful.

Lions in nature rise before dawn to hunt. They are raring to go — and hungry — at first light. It's an effective adaptation, since they are most energetic when their prey is vulnerable and sleepy. It is the lion's ambition to rise to a position of power within the extended family unit, aptly called a "pride." The human equivalent also rises before the sun comes up, is ravenous upon waking, and, after a hearty breakfast, is ready to conquer whatever goals he or she has set for that day. Lions burst with purposeful energy. Bright-eyed go-get-'em eagerness might be their defining characteristic. They face challenges head-on, with clear objectives and strategic plans for success. Life is like a straight line. It's about getting from point A to point B, by taking steps C, D, and E. With their analytical, organized minds, they're not easily lured by temptation; nor do they take big risks. In fact, Lions steer away from unnecessarily dangerous or troubling situations.

Lions assume leadership roles in groups. As introverts, they find it a bit lonely at the top. Completing a project or task gives them a profound feeling of accomplishment. When things go wrong, Lions take setbacks in stride and calmly readjust their strategy to get back on course. In business, their risks are calculated. No fly-by-night crazy winging it for *them*. Before Lions start a new venture, they have a business plan with dotted i's and crossed t's. And, not surprisingly, the venture is likely to be a success. Most CEOs and entrepreneurs are Lions.

Lions prefer to be around other people, but since they tire early in the evening—having been firing on all cylinders for hours already—they are unlikely to stay out late with friends. They're usually the first to leave a party, saying that they have to get up early for a meeting or to train for a marathon. In relationships, Lions tend to emphasize the positive and to be "fixers"— trying to solve problems instead of brooding and ruminating about them.

Since Lions prioritize their health, they're less likely to overindulge in junk food and alcohol. But they can use these substances to calm them down. Lions generally eat a healthy diet of lean proteins and high-fiber grains, fruits, and vegetables. Exercise presents another way for Lions to set and achieve goals, which the majority of them take advantage of. They are the mud runners and CrossFit enthusiasts of the world. Lions tend to have a low BMI and rate their overall life satisfaction as very high. Check out your perfect day on page 50.

BEAR

- **Four Key Personality Traits:** Cautiousness, extroversion, friendly and easy to talk to, open-minded
- **Four Key Behaviors:** Avoiding conflict, aspiring to be healthy, prioritizing happiness, taking comfort in the familiar
- **Sleep/Alertness Pattern:** Bears wake up in a daze after hitting the snooze button once or twice, start to feel tired by mid- to late evening, and sleep deeply but not as long as they'd like. Most alert: mid-morning into early afternoon. Most productive: late morning. Naps: Bears catch extra hours on the weekends, on the couch.

When bears in nature are not hibernating, they are diurnal—active in the day and restful at night. Grazers, they forage for food continually and will eat regardless of when they had their last salmon. They are playful and affectionate within the family unit, and form close friendships within their larger society.

Human Bears' sleep/wake patterns match up with the solar cycle, which is fortunate for them. With their high sleep drives, Bears would prefer to sleep for at least eight hours per night, if not longer. It takes them a couple of hours to feel fully awake in the morning. They are often hungry upon waking, and may be hungry all the time. If food is available, they'll probably eat it, even if it's not a meal or snack time. Their diet isn't particularly good or bad; they may or may not be dedicated exercisers. Bears self-report their general health as fair. If they make an effort with diet and exercise, they do so sporadically, with mixed results. Bears' BMI tends to be average to high.

In the professional realm, Bears are team players, balanced thinkers, worker bees, and middle managers with good people skills. Affable Type B personalities, Bears don't do drama often. They're not likely to scheme to get a colleague's job or to blame others for their mistakes. In school, they were solid students, and the same can-do attitude applies on the job. They aspire to do decent work and then go home and put their feet up. Risk-averse, Bears are unlikely to put themselves out on a limb professionally or personally, unless they think they are a shoe-in or something happens organically.

Bears like to be around other people and grow restless and bored if they're alone for too long. At a party, the gregarious guy manning the bar or flipping burgers at the barbecue is probably a Bear. In their personal relationships, Bears can be easygoing to a fault. They tend to rate low in both emotional repair (the ability to "fix" problems) and clarity (understanding what's really going on). This can be frustrating for their partners, especially insightful Wolves and anxious Dolphins.

Bears don't have high highs and low lows. If they do get knocked off their even keel emotionally, it's a direct reaction to a real-life crisis. When the issue passes, so will their anxiety or depression. Their overall life satisfaction is good. Check out your perfect day on page 63.

WOLF

- **Four Key Personality Traits:** Impulsivity, pessimism, creativity, moodiness
- **Four Key Behaviors:** Taking risks, prioritizing pleasure, seeking novelty, reacting with emotional intensity
- **Sleep/Alertness Pattern:** Wolves have difficulty waking up before 9:00 a.m. (they do it, but they're not happy about it), are groggy until

midday, and don't feel tired until midnight or later. Most alert: 7:00 p.m. Most productive: late morning and late evening. Naps: Tempting, but if a Wolf sleeps during the day, he won't fall asleep at night. It's just not worth it.

In nature, wolves come alive when the sun goes down. As night stalkers, they hunt in a pack and are creative, ferocious, and cunning. They see all the angles, even in the dark.

Wolf chronotypes are often proud of their fearlessness. They are the most impulsive and spontaneous of all the chronotypes, willingly flinging themselves into risky situations. They are always on the hunt for new experiences and sensations. To that end, Wolves tend to have a higher than average number of sexual partners in their lifetime. Although they're not hungry at all in the morning, they become ravenous after dark. Wolves are likely to drink soda and alcohol and to eat high-fat, high-sugar food, often late at night while standing in front of the fridge. Not surprisingly, Wolves' BMI is average to high. Due to their eating schedules and poor choices, they're more likely to have obesity-related diseases like diabetes, a danger they share with Bears.

Wolves are completely comfortable being alone and might come across as standoffish. But they are also extroverted and love a good party. When in the right mood, they can be the center of attention and are often the last to leave a gathering.

It's not all wine and roses, however. Wolves have intense feelings and can't usually suppress them. Their emotional reactivity is high, meaning they fly off the handle easily, a trait that can be stressful for them and their families and friends.

The Wolf mind is insightful and intuitive. Right-brained Wolves are always connecting the dots in strange and inventive ways. They tend to work and thrive in creative fields—the arts, medicine, publishing, and technology. Creativity does not necessarily correlate with academic achievement, however. In school, Wolves were likely to have been slackers or coasters. The phrase "not working to potential" might've come up in teachers' comments.

Wolves are the iconoclasts of the world. Their chronorhythm is out of step with 80 percent of the population, and not in a good way like Lions'.

The stress of being out of sync — and the perception that they are "lazy" — exacerbates a Wolf's already intense emotions. Wolves tend to suffer from mood disorders like depression and anxiety more often than do other types. Their drug and alcohol use might be as much self-medicating as it is novelty-seeking. Wolves are more likely to be addicts than other types. Their self-reported life satisfaction tends to be low unless they are married to another Wolf, and then their satisfaction can be high. Check out your perfect day on page 77.

THE TEMPERATURE TEST

Still not sure of your chronotype? There's a biological way to determine your kind, and all it requires is a little bit of commitment and a digital thermometer.

Your hypothalamus regulates your temperature within a pretty narrow range of 96.8 to 100.4 degrees Fahrenheit. A nightly drop in "core temperature" (vital organs) as opposed to "shell temperature" (skin and muscles) is a signal to the body to feel sleepy; a morning rise in temperature signals that it's time to wake up. Your chronotype determines when you'll hit your temperature high and low points.

To take the Temperature Test, record your temperature every hour between 5:00 p.m. and bedtime. When does your temperature start to rise? When does it start to drop? The rise in temperature won't be dramatic. It could be just a few tenths of a degree (this is where digital thermometers come in). Chart your temperature for three days to collect sufficient data.

Dolphin: This will not work for you. Dolphins aren't like other types: Their core temperature goes up at night. It's one of the reasons they have difficulty falling asleep.

Lion: Temperature starts to drop at **7:00 p.m.**

Bear: Temperature starts to drop at **9:00 p.m.**

Wolf: Temperature starts to drop at **10:00 p.m.**

Still can't tell which chronotype you are? Pick the one that seems to fit you 80 percent of the time, then ask your partner, a parent, or a very close friend (we often do not see ourselves as we are), take that person's word for it that you are who he or she says you are, and just go for it. Follow the daily schedule outlined in the upcoming chapters for your type. After a week, if it doesn't fit you, or if it does, then you'll know which chronotype you really are.

Before I get into the specific schedules, I want to send up a couple of alerts:

- **The schedules are optimal.** In some cases, optimal might not be practical. If parts of the schedule are in complete conflict with your life, don't worry about it. Just do what you can, and you'll see benefits. Any changes that get you closer to perfect bio-time are positive.
- **The schedules are biological.** They represent what your body wants. They might not represent what your mind wants. Most of us don't even like the idea of a schedule. Our minds want freedom, and nothing is as limiting as a schedule. Instead of thinking of it as limiting, think of it as what you need to do for limitless possibility. Remind yourself that real freedom is boundless energy, losing the burden of excess weight, having better communications and a supercharged immune system. If freedom is attainable by shifting dinner up or back by an hour, or exercising later or earlier in the day, it's a small price to pay.

CHAPTER 2

A Perfect Day in the Life of a Dolphin

Stephanie,[1] a fifty-three-year-old teacher from New York City and mother to a college-aged son, arrived for her appointment with me early. By chance, I observed her in the lobby of the building. Instead of scrolling through emails or reading a book, Stephanie paced the lobby. Although Dolphins like Stephanie are chronically tired, they are also wired with nervous energy.

The description "tired and wired" fit her well. When she sat down for our consultation, she described herself as exhausted. Fidgeting with a pen, she readjusted her position often. She couldn't settle down. I asked her how much sleep she got on average per night.

"I'd be thrilled to get six solid hours, but it's usually more like four or five," she said. "It's hard to say, because I wake up throughout the night and I'm not sure when, or if, I fall asleep again." How often did she wake up? "Probably five or six times. But, again, I can't really say. My husband could probably tell you."

Stephanie's husband was the reason she came to see me. "I've always had trouble sleeping, even as a kid," she said. "But since menopause, it's been a lot worse, and my husband has reached his limit with my tossing and turning. I'm driving him nuts." She described their close relationship. "He's my best friend. He might be the only person who really knows me, apart from our son." Dolphins tend to be neurotic and private. But once the intimacy wall is breached and all of their quirks come out, they can form extremely close, loyal relationships.

Stephanie walked me through a typical day in her life. "I stop trying to sleep at around six in the morning, sometimes earlier. I lie there for a while and hope I fall back to sleep, but there's no point. So I get up, take a shower, and have breakfast. I eat a big breakfast—a bowl of cereal and milk with banana slices plus a bagel half with cream cheese and a muffin or something sweet with my coffee."

Considering how much she eats every morning—the cereal/bagel/muffin combo is upwards of 800 calories of mostly carbohydrates—you might think Stephanie would be heavy. But she is thin, wiry, a typical Dolphin body type.

I asked about lunch and dinner. She said, "I teach math to middle school kids, and have office hours at lunchtime. I often lose track of the time and forget to eat. I typically have a quick dinner at home with my husband, or I grab something on my way home from school." So food is not high on Stephanie's priorities list. Apart from breakfast, her meals are afterthoughts. Dolphins do, however, find themselves snacking during the day. I think it's an effort to self-medicate (calm themselves down) with foods that give them comfort and a hit of serotonin (carbs).

Dolphins tend to drag, so I asked about her alertness and concentration levels. Stephanie said, "It's hard to concentrate in the morning because I'm really tired, but I seem to do better by the afternoon. It's like the day doesn't really begin until one or two o'clock. During the evenings when I correct tests and do my home budgeting, though, I can really lock in and concentrate. I feel most alert when I finish everything around eight or nine at night," she said.

I asked Stephanie what she does with her evening burst of nervous energy. "I do a quick lap online—email, Facebook, shopping. I have a snack. I might get caught up watching a movie or TV show. I have been known to start cleaning out a closet or doing laundry at 11:00 p.m. I just see a chore that needs doing and have to get into it or it bugs me. I finally get in bed around midnight or a bit later. I use a mouth guard to prevent grinding my teeth. Also, I use earplugs and an eye mask," she said. "I should pass out when my head hits the pillow, considering how little sleep I had the night before, but my mind and heart start racing. And then the nightly wrestling match begins. I try to will myself to sleep, but my mind won't switch off."

Stephanie came to me to help her get both more and better sleep. Her insomnia was affecting her relationship with her husband and the quality of her life. Insomnia is thought to be a nighttime complaint, but it's actually a twenty-four-hour problem. Stephanie operates at a low level of energy and efficiency all day long. She doesn't take much joy in her routine, describing it as a bit of a slog.

Although she doesn't realize it, Stephanie and other Dolphins like her have upside-down biochemical patterns.

Cortisol is the fight-or-flight hormone, released by the adrenal gland when the body is under stress — not the biochemical response you want when you're trying to relax and rest. For every other chronotype, cortisol levels drop at night. But for Dolphins, cortisol levels are *elevated* at night.

It'd be logical to assume that cortisol levels go up at bedtime for insomniacs because of their sleep problems. It's absolutely true that insomniacs do get anxious about their long, restless night ahead. But if anxiety were the only cause, their cortisol level should drop when they finally pass out. It doesn't. Researchers at the University of Goettingen, Germany, tested[2] the plasma cortisol secretions of seven severe insomniacs over the course of the entire night. The subjects' cortisol levels remained elevated, even when they were asleep. The higher their cortisol level at bedtime, the more frequently they woke up throughout the night.

At wake time, cortisol production in Lions, Bears, and Wolves goes up to get them moving. But in Dolphins? Their cortisol levels are at their lowest in the morning. Researchers at the University of Luebeck, Germany, tested[3] the salivary cortisol in fourteen insomniacs and fifteen healthy, normal sleepers. The insomniacs' morning cortisol levels were significantly lower than were those of the control group. The lower their salivary cortisol, the lower their self-reported sleep quality.

Dolphins' body temperature takes longer to drop at night, and their cardiovascular rhythms are flipped. For Lions, Bears, and Wolves, blood pressure goes down at night as their bodies shift into a hypoaroused (meaning underactive) state of relaxation. At night, Dolphins shift into a hyperaroused (meaning overactive) state, with elevated blood pressure. In a 2015 Mayo Clinic study, sleep-deprived subjects' systolic and diastolic blood pressure markers rose at bedtime. The numbers didn't go down when subjects were sleeping, which, as you can well imagine, negatively affected the quality of their sleep.

When Stephanie says she feels wide-awake when her head hits the pillow, she's absolutely right. Despite her exhaustion, her body is on high alert. Dolphins' biology makes it hard for them to relax.

What about the Dolphin brain? In a normal sleeper's brain, regions associated with unfocused mental wanderings are active *only* during waking hours, when the sleeper isn't focused on a specific task. In contrast, the brain's wandering-mind regions *light up when a Dolphin is sleeping.* Dolphins' night dreams are more like daydreams. When patients like Stephanie tell me that they have no idea whether they sleep at all, it's because their minds are wandering when they should be resting and consolidating.

During the day, their brains are similarly hyperactive. In 2013, researchers at the University of California–San Diego[4] gave insomniacs and good sleepers memory tests while scanning their brains with MRI machines to measure cognitive performance. As the questions got harder, the good sleepers' wandering-mind regions dialed down as their working-memory regions lit up. Not so for the insomniacs. Their wandering-mind regions stayed switched on, even when they needed to concentrate harder on the memory tests. The insomniacs' cognitive performance was on par with the good sleepers', so their unusual brain activity pattern didn't prevent them from getting the answers right. But the MRI results do shed light on why Dolphins often complain about poor focus despite performing at a high level.

Dolphins really do swim in different waters than the rest of humanity. Despite their upside-down physiology and personality quirks, they can achieve and excel in a Bear's world and do remarkable things. Stephanie, for example, teaches algebra to teenagers, something 99.9 percent of the population couldn't imagine undertaking. And yet what could she accomplish if she weren't so tired (from insomnia) and wired (from a hyper brain, elevated cortisol, and high blood pressure)?

Most of my patients fall into the Dolphin category. I've worked closely with hundreds of them and have watched them transform their lives simply by switching around the timing of the things they do every day. I set very simple and straightforward goals for my Dolphin patients:

- **Increase energy in the early hours to make better use of morning hours.**
- **Decrease anxiety in the evening for a more restful night.**

Dolphins' physiology presents challenges, no question. They also have psychological factors to contend with if they want to reach their chronorhythmic goals.

- **Unrealistic expectations.** Insomniacs need to lose the illusion that if they could get eight hours a night, all of their problems would go away. It is not in a Dolphin's biology to get eight consecutive hours of deep sleep night after night. I can help them get six hours of sleep regularly, and that will provide them with all the physical restoration and brain consolidation they need.
- **Inconsistency.** The power of when comes into full force with *consistency.* Dolphins can use their neuroticism in a positive way by fully committing to my recommended changes for an entire week. I urge patients to make their chronorhythm (below) a new obsession and compulsion; to set alarms to remind them to do each activity at the same time each day—especially wake and sleep times. If they can manage that, they'll notice immediate improvements.

REALITY CHECK

The following schedule is how you'd organize your day in a perfect world. But real life is not perfect. Due to social or work situations that are out of your control, you might not be able to follow the schedule to the letter.

That's okay.

The worst thing you can do is say, "If I can't do X, Y, or Z at exactly the right time, the whole thing is thrown off, so forget it." *Any* **changes will result in improvements to your health and happiness. It's not an all-or-nothing proposition. Ideally, you could do it all. Practically, you might not be able to. So do what you can now. Over time, as you notice positive changes, you might find that you can do a little more.**

THE DOLPHIN'S CHRONORHYTHM

6:30 A.M.

Typical: As Stephanie says, "I'm too tired to get up, and too wired to fall back to sleep."

Optimal: Get up and move. Your blood pressure, body temperature, and cortisol levels are low, so use exercise to turn them up. It might be the last thing you want to do when sapped by sleep inertia—that groggy, just-woke-up feeling. But do it anyway. I tell patients to roll out of bed, right onto the floor, and do a hundred crunches. Then flip over and do twenty push-ups. In just five minutes, your heart rate will go up. Muscle stress (a good kind) raises your cortisol level. During the first five minutes of your day, flip your physiology from exhausted to energized. Ideally, you'd get a twenty-five-minute workout in, but even a few minutes of cardio will help. If possible, get five to fifteen minutes of direct sunlight to activate your SCN during your workout or cool-down.

7:20 A.M. TO 9:00 A.M.

Typical: "I drag myself into the shower and then eat cereal and a bagel and get myself to school for my first class."

Optimal: Jump-start with a cool shower and a high-protein breakfast. If you have done a few cardio moves as recommended, by all means, rinse off. Since a hot shower might lower core body temperature (it sends blood to the extremities), take a cool shower instead, as cool as you can stand it to rush the blood into your vital organs, raise core temperature, and trigger "I'm awake now" hormonal secretions. This is also a great time for a one-minute meditation. Let the water run over your head, and think of nothing for sixty seconds. It will bring you into the "now" and help you focus. I use this technique most mornings, and it is wonderful.

Before you take a single bite of breakfast, drink a large glass of room-temperature water. Everyone is dehydrated after a night's rest, especially Dolphins, whose metabolism works overtime overnight. You need to replenish depleted cells with fluid and the right nutrients from food. Although you might crave a bagel or a bowl of sugary cereal for quick energy, *morning is the wrong time to eat carbs*. Carbs increase the production of serotonin, the "comfort" hormone. You might need the food hug after a bad night, but it's the opposite of what you need hormonally. When serotonin levels go up, cortisol levels go down, relaxing you. A bagel will hit your metabolism like a tranquilizer dart. Instead, **eat protein in the morning** to boost cell recovery and fuel muscles: eggs and bacon, yogurt, a protein shake, or a small serving of oatmeal with seeds and nuts.

9:30 A.M. TO 12:00 P.M.

Typical: "Like I'm in a fog. I can't make myself feel more alert. I can barely concentrate."

Optimal: Think. You *can* burn off the fog of sleep inertia with exercise, a cool shower, and a plate of eggs. If you drink coffee, now would be a good time to use caffeine wisely, to deactivate the sleepy neurochemicals. Have one cup only. Two cups will make your jittery. If you are already a big caffeine addict, do not immediately drop off to one cup. Check out my video on caffeine fading at www.thepowerofwhen.com.

Since you're still gaining alertness, it's not the best time to try to zero in and focus. Instead, use morning as a great time to brainstorm. Let your mind wander. See what brilliant ideas bubble up. When you're slightly tired, your hyperactive, creative mind is primed to do what it does best: connect the dots, no matter how disparate and misaligned they seem. If you are into journaling or jotting down big-picture ideas, this is the perfect time to do so. It's what many Dolphins do at night while trying to go to sleep. But it's much better for them to do it in the morning—or anytime before nightfall, if possible.

12:00 P.M. TO 1:00 P.M.

Typical: "If I get distracted, I might forget to eat lunch."

Optimal: Eat something! Dolphins tend to have a wiry, lean body type. Neither chronic dieters nor foodies, they eat to live, and sometimes they can't be bothered or they forget, especially when they fall down a rabbit hole (warning: the Internet is full of them). Set an alert on your phone to remind you to eat something at 1:00 p.m. every day. Replenish yourself with nutrients that will fuel your body and brain—one-third carbs, one-third protein, one-third fat—and will keep your state of hyperarousal on an even keel. Some suggestions: a sandwich, a burrito, soup and a salad. Always drink plenty of water, too. If you had coffee earlier, don't have more with lunch. Too much caffeine will not energize you. It'll only make you jittery, could decrease your appetite, and might keep you awake at bedtime (yes, even many hours later).

1:00 P.M. TO 4:00 P.M.

Typical: "Early afternoons are a struggle. I would love to close my eyes and take a nap. Sometimes, if I have time, I drop my head on my desk and close my eyes."

Optimal: Recharge. Do not nap! Napping lowers the buildup of sleep pressure, making it harder to pass out at bedtime, already a challenge for you. Your goal is to improve the duration and quality of sleep *at night*. Taking an afternoon nap is self-sabotage. Do not drink coffee! No caffeine of any kind for Dolphins after 1:00 p.m. If your energy levels drop midday, recharge by being active. Whenever you lose steam, I want you to train yourself to think, "Exercise and sunlight." Raise your blood pressure, heart rate, and cortisol levels by working large muscle groups. You don't even have to break a sweat. Just take a walk around the block, around the office, wherever, ideally outside to soak up more sunlight.

4:00 P.M. TO 6:00 P.M.

Typical: "Too much coffee! I don't know if I can't concentrate from being tired or jacked-up on caffeine."

Optimal: Scale the wall. While the Bears and Lions around you are starting to peter out, your cortisol level is on the rise, making you as alert as you've felt all day, especially if you've kept the carbs to a minimum and if you took an afternoon stroll. Let your inner neurotic emerge. Obsess over a project. Do the heavy lifting intellectually and mentally. If you had a thread of an idea earlier during your off-peak morning brainstorming session, it's now time to reel it in. If you work in an office environment, close the door to your office or get some kind of privacy (perhaps by putting on some imaginary blinders), and apply your afternoon peak alertness to figuring out the details of a specific project or task.

6:00 P.M. TO 7:00 P.M.

Typical: "I get hungry now from skipping lunch, and crave something quick and ready to eat. I'd be happy to grab a slice of pizza for dinner every night."

Optimal: Be alone. Don't eat yet. Since you set an alarm at lunch to remind you to eat, your hunger is manageable. Instead, take the post-work break for strategic downtime. Schedule fifteen to thirty minutes for quiet alone time to decompress. Your hyperactive mind will become increasingly anxious as the night wears on and cortisol levels rise. Starting the evening with quiet alone time can ward off or lessen those hormonal and emotional reactions.

Some Dolphins might do meditation or yoga. Others take a counterintuitive approach to quieting anxious thoughts by sitting alone in a quiet place and letting themselves ruminate on worst-case scenarios for a limited time. The purpose is to habituate yourself to anxiety by "going there" daily, and to save random worries that pop up throughout the day for that one time period only. Eventually, if you do this consistently, anxiety amplitude — the intensity of your worry — will decrease, as will the amount of time you spend worrying. This strategy is particularly useful for insomniacs, but only if they really commit to this practice at the same time every day. Set an alert or alarm to remind yourself that it's time to "go there."

To reap the benefits of a retrained brain and reset chronorhythm, you have to be consistent.

Most patients are surprised that they can't ruminate darkly for the full fifteen minutes. If you're done with it after five minutes, spend the rest of your alone time counting breaths up to ten, then backward to one, and repeat. Go to www.thepowerofwhen.com and get a free video step-by-step of this breathing technique.

6:30 P.M. TO 8:00 P.M.

Typical: "After dinner, it's go time, like I'm fully awake and have energy. I run errands or start organizing things around the house or on the computer."

Optimal: Prepare and eat dinner. *Now* is the time for carbs. Lean body-type Dolphins are not usually on a diet. Have a big bowl of mac and cheese or a baked potato. It's called comfort food for a reason. Your serotonin level will go up and your cortisol level will go down, calming your hyperaroused body and hyperactive mind. If there is anything you need to discuss with your partner or family that might be upsetting or may cause consternation, do it while you eat. The serotonin level uptick will serve as a buffer for tense or anxious feelings.

8:00 P.M. TO 8:30 P.M.

Typical: "I get a lot done, or try to. I might set out to do one thing and get distracted by something else, especially online. There is always a chore or thing I have to do."

Optimal: Have sex, either with a partner or by yourself. It might seem odd to do it at 8:00 p.m., but post-dinner, pre-bedtime sex serves a couple of purposes for you. Not only does sex have soothing physical and emotional benefits—including a blast of oxytocin, the relaxing "love hormone"—but it will help you redefine what "bed" means to you. If you engage in a positive, loving experience there, one that doesn't immediately precede the dread and anxiety of trying to fall asleep, you'll reinforce positive associations with bed and condition yourself to think of it as a fun, not scary, place. If you usually have sex immediately before sleep to relax, the effort could backfire. The exertion will feed into your anxiety, reinforcing the negative association with the thought "light switch off, brain switch into overdrive."

8:30 P.M. TO 10:30 P.M.

Typical: "Since I didn't sleep well last night, I go to bed early to catch up. But it doesn't work. When I lie down, my brain goes haywire. I think of stuff I need to do or would like to do. I might scroll through Facebook or finish watching a movie on my phone to get my mind off the insomnia."

Optimal: Power down. The post-dinner hours are all about relaxation. Direct your evening energy surge toward something purposeful but nonengaging to soothe and quiet your mind. Watch TV with your family or go out to a movie. Take a walk to get ice cream (more carbs!). Go ahead and clean out a drawer or whatever chore or task takes hold of your mind, but do it with the awareness that you have to stop at 10:30, no matter what. If you meet a friend for a drink or have a glass of wine, make sure your last swallow is by 9:00 p.m. Alcohol can disrupt sleep, and you need to give your body adequate time to metabolize the alcohol out of your system by bedtime.

10:30 P.M. TO 11:30 P.M.

Typical: "Still lying in bed, awake. I start to get frustrated and get on a bad loop of being anxious about not falling asleep, which only makes it harder to do. Or I'll think of something important and wake up my husband to talk about it. This does not go well."

Optimal: Power down. Using electronics at night makes it harder to fall asleep. The blue wavelength light emissions suppress melatonin secretions. Avoid any blue light by turning off all screens, including your phone, at 10:30 p.m. I call it the Power-Down Hour. If you must watch

TV, dim the brightness and make sure the picture is at least ten feet away from your eyes. In fact, dim all the lights in the house to stimulate the production and release of melatonin. I recommend special lightbulbs that have been created to filter out blue light at night. If you are interested, go to www.thepowerofwhen.com and see my video on these amazing discoveries (they're great for kids as well). Meanwhile, shift your focus to nonscreen activities that lower cortisol levels and blood pressure. If cleaning or organizing engages your brain—and I know it does—you have to stop that now, too. A great idea is to take a hot shower or bath to help lower your core body temperature. Engage in quiet, casual conversation. Cuddle. Meditate. Do some low-intensity stretching. What bores you to death? Do it now. What gets your juices flowing? Avoid it! Your wandering mind should not be checking emails that could excite or irritate you or scrolling through Facebook and clicking cool links. I discourage insomniacs from reading biographies or memoirs at night. Among my patients, nonfiction tends to be more mentally engaging than fiction. A novel is a safer choice. Even better, read something tedious and boring, like a computer instruction manual. Just don't turn it on!

11:30 P.M.

Typical: "Still lying awake."

Optimal: Go to bed. Dolphins should not get in bed until now. In fact, except for 8:00 p.m. sex, don't spend *any* time in bed before now. Don't hang out or watch TV there. Don't read in bed. You must learn to associate bed with sex or sleep *only.*

Once in bed, try progressive muscle relaxation (go to www.thepower ofwhen.com to watch video instructions) or counting backward from 300 by threes. If you are not asleep in twenty minutes, get up and sit in a chair in the dark for fifteen minutes before returning to bed to try again. Repeat these twenty-minutes-in, fifteen-minutes-out cycles. The strategy is called "stimulus control." The concept is to avoid the buildup of anxiety from just lying there. You might have a few bad nights using this strategy, but eventually it'll help lower your anxiety and blood pressure in bed and will yield more continuous quality rest.

12:30 A.M. TO 2:30 A.M.

Typical: "Tossing and turning. My anxiety is ramping up. I look at the clock and calculate how many hours of sleep I might get if I pass out in ten minutes or twenty. My whole body feels tense."

Optimal: Enter Phase One. If you follow the chronorhythm I've outlined for you and consistently practice the strategies, you will be able to fall asleep within thirty minutes of bedtime. This will take some time to accomplish (maybe a week to ten days of consistency). The first two hours of sleep are the most important for you. During Phase One, your body is physically restored. All the tension of the day is released from your muscles and brain to repair and rebuild on a cellular level from bones to skin. During the first week on this schedule, insomniacs won't always be able to pass out within thirty minutes, but don't give up. Stick with it. I've helped hundreds of patients retrain their bodies this way. Don't watch the clock. This will only frustrate you and make you do the "mental math," as Stephanie described above. If you are practicing stimulus control, use the stopwatch function on your phone, but don't check the time, just guess it.

2:30 A.M. TO 4:30 A.M.

Typical: "If I'm asleep, it's like skimming the surface. I wake up repeatedly and don't know if I fall back under."

Optimal: Enter Phase Two. Not a lot goes on during the middle portion of the night. Phase Two is uncomplicated sleep. If you do wake up momentarily, don't let it upset you. **Wake-ups are completely normal.** We all wake for a few seconds at the end of every ninety-minute sleep cycle before beginning another one. Deep sleepers don't remember doing it, but it happens. Dolphins are light sleepers and more prone to wake-ups than most. **Change your perspective on those wake-ups.** They're to be expected—a healthy part of sleep. If you see them as normal, you won't obsess about having them or view them as a sleep failure. Anxiety will be lessened and the wake-up periods will become shorter.

4:30 A.M. TO 6:30 A.M.

Typical: "I rarely see 4:00 a.m. on my clock. Even I am asleep then."

Optimal: Enter Phase Three. During the final portion of the night, you get the bulk of rapid eye movement (REM) sleep, when you consolidate memory and effectively clear out the cobwebs in your brain. For short-sleeping Dolphins, two hours of Phase Three is a great goal.

6:30 A.M. TO 7:00 A.M.

Typical: "I wake up tired and vow to go to bed early tonight, if I can just get through the day on fumes."

Optimal: Wake up refreshed. If Dolphins can get a solid six hours, with adequate time in all three phases of sleep, their bodies and brains are rested and fit to face the challenges of the day.

On weekends, get up at your scheduled wake time, even if you think you can sleep longer. Sleeping in is a trap for two reasons:

One: It won't do you any good. You need deep, physically restorative Phase One sleep (delta sleep), which you can get only at the beginning of the night. Extending the last third of sleep won't make you feel any better.

And two: Sleeping in will dismantle your carefully constructed chronorhythm. Remember, Dolphins thrive on consistency, which includes a consistent wake time every day, even on weekends and vacations. Otherwise, you'll push your whole day out of sync and you won't be able to fall asleep at midnight, setting off a chain reaction that you know all too well of not sleeping at night and dragging all day (this is called social jet lag). If you rise consistently and are active throughout the day, taking periodic walks, I guarantee that the quality of your sleep will be significantly improved and it will boost your day-to-day energy and sharpness far more than an extra hour of REM sleep on Sunday, which, by the way, is usually lighter sleep and not particularly refreshing.

EASY DOES IT

It seems like a lot of change, and it is. But if you adjust your schedule slowly, making one or two small shifts per week, you'll be able to incorporate and internalize them into your life seamlessly. You'll notice significant improvements in the overall quality of your life in just one month if you make steady changes week by week.

Week One

Establish a consistent wake time and bedtime.

Elevate your heart rate upon waking with exercise.

Wean yourself off caffeine after 1:00 p.m. It doesn't help you wake up, and it can affect your ability to fall asleep at night. Switch to herbal tea.

Check out the videos on www.thepowerofwhen.com.

Week Two

Continue with previous week's changes.

Eat protein for breakfast, a balanced lunch, and 60 percent carbs for dinner.

Cool showers in the morning and/or hot baths at night.

Week Three

Continue with previous weeks' changes.

Take an afternoon walk.

Have your important, intense, and heavy conversations in the late afternoon/early evening.

Week Four

Continue with previous weeks' changes.

Practice stress-busting activities including pre-dinner meditation and post-dinner sex.

Power-Down Hour. Turn off all screens at 10:00 p.m.

Try stimulus control, or getting out of bed if you haven't fallen asleep in twenty minutes, sitting quietly in the dark for fifteen minutes, then returning to bed to try again.

Dolphin Daily Schedule

6:30 a.m.: Wake up, no snooze.

6:35 a.m.: Exercise on the floor of your bedroom or get dressed for a twenty-five-minute outdoor workout. If you work out indoors, try for ten minutes of direct sunlight during cooldown.

7:10 a.m.: Cool shower, including one-minute meditation.

7:30 a.m.: Breakfast, high-protein.

8:00 a.m.: Get dressed and organized.

8:30 a.m.: Out the door to work, or, for the self-employed, get right to it.

9:30 a.m. to 9:45 a.m.: Coffee break.

10:00 to noon: Creative thinking time. Daydream and journal for ideas. Make big-picture to-do lists, research, think.

Noon to 1:00 p.m.: Lunch. Do not skip!

1:00 p.m. to 4:00 p.m.: DO NOT NAP. Do not drink coffee! If you feel tired, take a walk—outside, if possible. Sunlight exposure will help.

4:00 p.m. to 6:00 p.m.: Peak alertness, most productive time. Tackle hard stuff.

6:00 p.m.: Fifteen minutes of alone time to decompress.

6:30 p.m.: Cook dinner, high-carbohydrate.

7:00 p.m. to 8:00 p.m.: While eating your meal, have any intense, demanding, or practical conversations with family and friends. The carbs will buffer anxiety.

8:00 p.m.: Sex, with a partner or by yourself.

8:30 p.m. to 10:30 p.m.: Afterglow. The post-orgasmic flow of relaxing hormones will prime you for sleep. Get things done at home or online, or watch TV.

10:30 p.m.: Turn off all screens and stop any mentally engaging activity. Read a novel, have light conversations. Take a hot shower or bath.

11:30 p.m.: Go to bed. Practice "stimulus control" to combat insomnia-related anxiety. If you find that you are not falling asleep for an hour or more, move your bedtime later by thirty minutes.

CHAPTER 3

A Perfect Day in the Life of a Lion

Lions don't often seek help about circadian rhythm issues. They hit the chronorhythmic lottery, with bodies and brains that are designed for achievement. Benjamin Franklin and scientific research agree: Early risers tend to be healthy and, since they're more likely to be bosses and leaders at work, wealthy.

What about wise? Lions' IQs, on average, are not higher than any other chronotypes'. They just make the most of their talents. Lions are the star students, the overachieving hand-raisers and extra-credit doers from your high school class.

I met Robert,[1] 28, a marketing executive in Boston, at a business luncheon. As usual, when people find out that I'm a psychologist who specializes in sleep, they tell me their bedtime stories.

Robert described a fairly typical Lion schedule. He wakes up at 5:00 a.m. every day. By 10:00 p.m., he can't keep his eyes open. Other clues — his high energy, trim shape, and choice of profession — indicated to me that he was a bona fide "member of the pride" (remember, a group of lions is called a "pride"). He is a go-getter, a natural leader, someone who gives you the impression upon first meeting that he will one day be the boss of everyone.

"The hours I keep are absolutely an advantage at work," he told me. "I get a lot done. I'm raring to go as soon as I wake up. My long-term plans and goals are on my mind almost constantly."

Early risers tend to be driven, focused, and goal-oriented in the professional realm. The current or former CEOs of Amazon, AOL, Apple, Avon, Cisco, Disney, General Motors, the *Huffington Post*, PayPal, PepsiCo, Starbucks, Unilever, Virgin America, and Yahoo!, to mention a few, have credited their pre-dawn wake time as one of the keys to their success, and they are **NOT** Lions. Facebook COO Sheryl Sandberg has said she leaves work early to be with her kids but gets up very early to send emails. I can only imagine how her non-Lion colleagues feel when they find pre-dawn emails from her in their in-box. You can practically hear the gulps of intimidation. Between 5:00 a.m. and 8:00 a.m., Lions exercise and catch up on reading or emails—or both at the same time. Organized multitaskers, Lions are born to get things done with brutal efficiency at a ridiculously (to other types) early hour.

Rising early is a function of their biology. Lions' cortisol level elevates and their melatonin wanes very early, typically at 3:30 to 4:00 a.m., which is why their eyes snap open before dawn. They don't struggle with sleep inertia, either. In a Lion's brain, the "white matter"—the fatty tissue in the frontal and temporal lobes of the corpus callosum that connects "gray matter" areas and allows nerve cells to communicate with each other— tends to be in excellent condition. Researchers at Germany's Aachen University compared[2] the brain structures of sixteen early risers, twenty-three late risers, and twenty intermediate types using advanced imaging technology, and found that the Lions' white matter was healthier than the Wolves'. When Lions' circadian rhythm signals "time to wake up," their brains respond obediently. Their cycles are also remarkably consistent. Lions rarely need alarm clocks to wake up at the same time each day, even on the weekends, even in a different time zone. They can suffer from jet lag severely because of this.

Except for crossing time zones, their consistent sleep schedule is a gift. A Canadian and French study[3] found that keeping regular mealtimes and hitting the hay early can ward off anxiety and depression and can even prevent bipolar and schizophrenic episodes in people suffering from those disorders. Numerous studies have confirmed that going to bed early is heart-healthy and BMI-friendly, which isn't too surprising, considering that late-night junk food consumption is a major cause of both arterial plaque buildup and weight gain. Lions are asleep when Wolves and Bears binge.

Lions come fully installed with all of these biological advantages. They're at the top of the chronotype food chain. But, as Robert confided,

"From a personal perspective, my daily schedule is not so great. My social life is on life support."

I asked him to elaborate. He said, "I'm awake before anyone else, and have two or three hours each morning to work, which is great since I'm focused on my career right now. But I'd like to start a family someday. It's nearly impossible for me to date and meet people organically—like at a party or a bar—if I'm comatose when everyone else is just getting started."

Robert described a recent date. The woman had to postpone their dinner until 9:00 p.m. "I'd ordinarily be getting ready for bed by then. At dinner, we had some wine, and it hit me like a ton of bricks. I tried to pay attention and seem excited to be there, but I yawned in her face at least three times. When I sent her a thank-you text the next day, she didn't reply. She must have thought I found her boring. Maybe she was boring. I just wasn't in the frame of mind to judge." He was off his bio-time for dating. (See "Fall in Love," page 94, for more about this.)

A Lion's mood and energy levels peak in the morning and steadily decline throughout the day. In a 2009 study,[4] researchers at the University of Liège in Belgium tested extreme early risers and extreme late sleepers' cognitive ability by examining their brains twice a day with an MRI scanner. An hour and a half after waking, both types were equally alert and proficient at the attention-related tasks they were given. But, ten and a half hours later, when the latesleepers got a second wind and showed more activity in the concentration areas of the brain, early risers hit a wall. The pertinent regions of the brain were essentially deactivated. It's all about how streamlined Lions are. Most of their systems are highly efficient, and that includes their sleep switches. When the signal to sleep turns on, they automatically shut down.

It's ironic that Robert's main complaint about being a Lion is its impact on his social life. He's wide-awake and active when 80 percent of the world is unconscious. Lions tend to be introverted (ruling the world isn't a social activity), but even the boss, or future boss, has a normal human need to share intimacies and to interact with other people.

"I know it's a cliché for people of my generation to have the fear of missing out. For me, it's not an irrational fear. I *know* I'm missing out on a

Lion

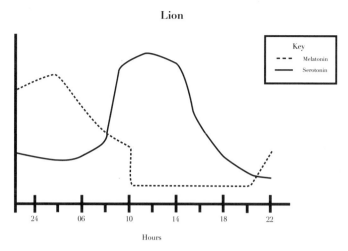

For Lions, "sleep hormone" melatonin levels start to drop around 4:00 a.m., causing an early wake time. "Happy hormone" serotonin levels peak in the mid-morning, putting Lions in a good mood.

lot of the fun that other people are having," said Robert. "I see the photos on Instagram and hear the stories about what happened at the party after I left. Real life is going on when I have to go to bed. And when I wake up, I'm the only man on Earth."

To make the power of when work, Lions have one simple goal:

• **To stretch their exceptional energy, positivity, and alertness further into the day so they don't power down so early at night.**

I've noticed two potential challenges for Lions in adapting to a new routine that will enable them to achieve their chronorhythmic goal:

• **A bumpy adjustment.** Lions are at the mercy of their efficiency. It's not easy for them to change their routines. What I'm suggesting in the daily schedule I've outlined below is not an overhaul but a gentle shifting of the same activities they'd ordinarily tackle. It might take a week to adjust to a new eating and exercise schedule. Lions might feel hungry or antsy in the meantime. But the discomfort won't last long, and the changes are necessary to squeeze another hour out of their evening for socializing.

- **Frustration.** Lion are accustomed to getting things done by sheer force of will, and they expect to see results. From a psychological perspective, they might be frustrated if the benefits don't occur immediately. Biology is not like the stock market. It doesn't change in a day, or even a week. I'd remind Lions that they will reach their bio-time goals, but just like in business, they have to climb the ladder one rung at a time. I suggest Lions mitigate frustration with determination. Pushing past the evening "wall" is worth a little effort.

REALITY CHECK

The following schedule is how you'd organize your day in a perfect world. But real life is not perfect. Due to social or work situations that are out of your control, you might not be able to follow the schedule to the letter.

That's okay.

The worst thing you can do is say, "If I can't do **X, Y,** or **Z** at exactly the right time, the whole thing is thrown off, so forget it." *Any* changes will result in improvements to your health and happiness. It's not an all-or-nothing proposition. Ideally, you could do it all. Practically, you might not be able to. So do what you can now. Over time, as you notice positive changes, you might find that you can do a little more.

THE LION'S CHRONORHYTHM

5:30 A.M. TO 6:00 A.M.

Typical: "As soon as my eyes open," said Robert, "I have to get moving. It's like I'm catapulted out of bed. I put on my sneakers and run a few miles, sometimes in the dark."

Optimal: Wake up, eat, and hydrate. With surging cortisol levels, you can't sit still, and naturally, you think of moving. But exercise increases cortisol levels and heart rate, making you even more alert. If you save exercise for the afternoon, however, you can get a much-needed energy boost when you're lagging. Instead of hitting the road when it's still dark out, go to the kitchen and eat breakfast, ideally within thirty minutes of waking. Drink two glasses of water after you eat. With a full stomach, you won't be tempted to exercise. Lions tend to make healthy choices, food-wise. I recommend a high-protein, low-carb breakfast for fuel at the start of a busy morning.

6:00 A.M. TO 7:00 A.M.

Typical: "I'm really hungry after my run. I'm hungry during my run! I run faster on the way back because I'm thinking about food. I eat first, then shower."

Optimal: Harness mental energy. Post-breakfast is an excellent time for Lions to sit down and contemplate the larger issues of life, like long-term career goals and the state of relationships. During those early-morning alone hours without distractions, your brain is primed for big-picture conceptual thinking. Make your to-do lists, and plan the day, week, months, and years ahead. Plot your path to world domination while the rest of the world is snoozing. Since you have the house (the block, the world) to yourself, you won't be bothered by others during a morning meditation practice. Taking a few minutes to not think about anything in particular can harness your energy as you burst forward into your day. So try it.

7:00 A.M. TO 7:30 A.M.

Typical: "I'm dressed for work and can't imagine sitting around at home, so I just head into the office. I'm there at least an hour before anyone else shows up. I use the time to clear my desk and get a jump on what's ahead."

Optimal: Have sex. After your cortisol and insulin morning jolt has leveled off, Lions should jump back in bed, but not for a quick nap. Morning testosterone in both men and women is highest within the first hour or two of waking, and sexual desire is strongest. It's the ideal time for Lions to have a sexual encounter, with a partner or alone. Be warned: If your partner is a Wolf, you might lose a hand if you initiate sex at this hour. Bear partners might appreciate the affectionate wake-up. Morning orgasms also give you an early dose of oxytocin, which will fill you with peace and calm for hours to come. If you have kids who need help getting ready for school, make it a quickie and save your epic encounters for the weekends. (For more about sex and timing, see "Have Sex" on page 107.)

7:30 A.M. TO 9:00 A.M.

Typical: "I'm on fire in the morning. I can really grind. If I have a report to write or something to research, I can knock it out in a few hours."

Optimal: Connect. If you've had a quiet morning of contemplation and sexual activity, you'll have taken the edge off your considerable energy. You'll still have plenty to spare, and you would do well to shine your morning light on others. Don't rush off to work before anyone else gets there. Instead, connect with other people. If you live with others, stick around at home, where your energy and positive outlook will be contagious and improve the moods of your family members. If you live alone, write emails, Skype with your parents, or have a breakfast date.

9:00 A.M. TO 10:00 A.M.

Typical: "Nose to the grindstone. Everyone else at work trickles in and takes a while to get up to speed. Meanwhile, I'm cranking."

Optimal: Impress and snack. Create opportunities to interact with colleagues and contacts first thing. You will be the star at a breakfast meeting. You won't be too hungry, since you've already eaten, but order something small, about 250 calories, that's 25 percent protein and 75 percent carbs (a yogurt with fruit, or a small bowl of oatmeal) and have coffee (three hours after waking, for a cognitive lift). A snack now will help you push your next meal into the early afternoon, which will extend your energy into the evening.

10:00 A.M. TO 12:00 P.M.

Typical: "I get hungry early in the afternoon, and I have to eat lunch. Ideally, I'd have lunch with colleagues, but they want to eat later and I just can't wait."

Optimal: Hold forth. If you are the boss or in a leadership role, call meetings for mid-morning, when you are best equipped, hormonally speaking, to make clear, strategic decisions. Mid-morning is your on-peak period, when your mind is sharp and analytical. The mental clarity lasts until noon, so squeeze the most out of this window of time to make your arguments, get your points across, solve problems, and find solutions. (If you have employees who are other chronotypes, learn how to work effectively with them in "Present Your Ideas," page 251.)

12:00 P.M. TO 1:00 P.M.

Typical: "When everyone else is having lunch, I start to slow down. It comes on pretty strong. After revving so fast for hours, I kind of deflate. But I get some coffee and power through."

Optimal: Eat lunch. Within an hour or two of having the midday meal, Lions (like Bears, Wolves, and Dolphins) experience an energy dip. If you ate on your old schedule, you'd be lagging earlier than the other types at a time when Bears are just entering their on-peak alertness. Not acceptable for Lions, who might be tempted to use coffee or energy drinks to regain their mojo. This is a physiological battle that you can't really win — unless you shift your afternoon meal to one hour later. If you had a nine o'clock snack, as suggested above, you could hold off eating lunch until noon, keeping insulin low until early afternoon and allowing you to maintain your morning energy for a while longer. If you can go outside for lunch, the exposure to sunlight will aid in your efforts to stay wide-awake. As for content, avoid heavy carbs that will make you sleepy. Shoot for a balanced meal of one-third protein, one-third carbs, and one-third healthy fats: a big salad with grilled chicken or salmon, a sandwich with one piece of bread, a fajita bowl with brown rice.

1:00 P.M. TO 5:00 P.M.

Typical: "I'm running on fumes. I've been awake for ten hours already, and by afternoon, I can't force myself to be 'up.' I might grab an energy drink or a protein bar to keep myself going."

Optimal: Float. Lions aren't known for their insight and creativity, but that might be because they squander their most innovative hours by forcing themselves to run on empty to solve strategic analytical problems. By afternoon, your analytical on-peak window is closed. Your problem-solving mojo is tapped out. Being off-peak is not necessarily a bad thing, as you have probably long believed. When Lions are tired and mentally fuzzy, they come into their creative and insightful powers. Stop trying to stay alert and on point. If you have freedom at work, this is the time to go purposely off point and think outside the box. Brainstorming meetings could yield innovative ideas.

Journaling is an excellent way to let your creative mind take over. Every afternoon, when you get a fifteen-minute break from work demands, take out an actual paper notebook and an actual ink pen, and doodle, scribble, or draw pictures or jot down whatever ideas float through your head. Turn your fuzzy focus to a particular subject, like your career or a relationship, but don't narrow the focus. Allow your thoughts to just drift and present themselves. Who knows what brilliant notions might appear? For a video about how to start a journaling practice, go to www.thepowerofwhen.com.

5:00 P.M. TO 6:00 P.M.

Typical: "I'm done. Nothing left in the tank. I'm easily annoyed due to tiredness, and I'm hungry. Of course, no one else is hungry, and I wind up eating alone."

Optimal: Exercise. Eating at this hour means a rise and fall of insulin, which will make you sleepier than you are already. Instead of eating at 5:00 p.m., exercise instead. Lions have a tendency to exercise at dawn because they're up and have nothing else to do. But if you can hold off working out until the evening, you'll get an energy boost by raising blood pressure, heart rate, and cortisol level when you really need it. Also, exercising when your body temperature is higher than first thing in the morning reduces the risk of injury—a major concern for health-conscious Lions. If the weather permits, catch the last rays of sunlight by exercising outdoors. If you shower afterward, make it another cool shower. When your core body temperature drops, as it does every late afternoon for you, you'll start to feel sleepy. Exercise followed by a cool shower will keep your core temperature up.

6:00 P.M. TO 7:30 P.M.

Typical: "So now my friends are ready to hang out. I'm not exactly the life of the party with my mood and energy rolling downhill. So I have a drink or two to change my frame of mind."

Optimal: Have dinner and one drink. A 6:00 p.m. dinner date is completely reasonable. Since you have had a later lunch and a post-work workout, you can hold off until now for dinner and join your Bear friends for a meal. Avoid carbohydrates, which will elevate "comfort hormone" serotonin level and decrease your already dwindling cortisol level. A plate of pasta at 6:30 p.m. would affect you like a powerful sleeping pill. To prolong energy, eat protein for your final meal, or eat light to keep your blood sugar low and avoid a crash.

In good conscience, I can't advise anyone to start happy hour at 4:00 p.m., but that is when a Lion's metabolism best tolerates alcohol. If you start drinking at dinnertime, you can handle one or two glasses without feeling flattened. But do not drink after 7:30 p.m., or your body won't be able to metabolize the alcohol in your system before bed, and your sleep could be poor in quality or disrupted.

7:00 P.M. TO 10:00 P.M.

Typical: "Beyond done. I've totally hit a wall. I've been awake for fifteen hours and every cell in my body is telling me to go to sleep."

Optimal: Live it up. Your lean and efficient sleep system would be telling you that it's time for bed now, but because you have subtly shifted your eating and exercise schedule to make sleep pressure less intense, you gain an hour or two of alertness before insulin, cortisol level, and blood pressure drop. Enjoy it! Don't rely on coffee or alcohol to keep you going. They won't work, and could ruin your sleep, which Lions need to conquer the world tomorrow.

If you're having a quiet night of relaxing at home, you can still socialize with family and friends online or on the phone. You bought yourself an extra hour of alertness for human interaction, so make the most of this time to connect and nurture your soul.

10:00 P.M.

Typical: "I'm down for the count."

Optimal: Downshift. I'd advise Lions to be in their home environment by 10:00 p.m. Prepare for 10:30 bedtime by turning off all blue wavelength light from phones, tablets, and monitors, or use the special lightbulbs I recommend at www.thepowerofwhen.com. Computer screens will suppress melatonin secretions and delay the onset of sleep—yes, even for Lions. It's okay to watch TV at this hour, as long as the monitor is at least ten feet away from your eyeballs.

10:30 P.M. TO 1:30 A.M.

Typical: "Completely out. A bomb could go off and I wouldn't know it."

Optimal: Enter Phase One. Since you are pushing your limits to stay up later than usual, you should fall immediately into physically restorative rest. You brain waves are slower and deeper earlier in the night than those of other chronotypes. You even sleep more efficiently!

1:30 A.M. TO 3:30 A.M.

Typical: "Dead to the world."

Optimal: Enter Phase Two. During the middle portion of the night, you will have uncomplicated rest.

3:30 A.M. TO 5:30 A.M.

Typical: "Still out."

Optimal: Enter Phase Three. During the last third of the night, you get the bulk of REM sleep when memory consolidation occurs. Wake up at your usual time refreshed and ready to conquer the world again.

EASY DOES IT

It seems like a lot of change, and it is. But if you adjust your schedule slowly, making one or two small adjustments per week, you'll be able to incorporate and internalize them into your life seamlessly. You'll notice significant improvements in the overall quality of your life in just one month.

Week One

Eat breakfast within thirty minutes of waking up.
Exercise in the early evening, not the early morning.
For a video about the timing of sun exposure and exercise, go to www.thepowerofwhen.com.

Week Two

Continue with previous week's changes.
Try making connections in the morning at home or at breakfast meetings (you'll have a snack) instead of pushing all interactions into the evening hours.
Shift lunch to 12:00 p.m.

Week Three

Continue with previous weeks' changes.
Shift dinner to 6:00 p.m.
Drink alcohol after 7:30 p.m. only one or two nights per week.

Week Four

Continue with previous weeks' changes.
Schedule important strategic meetings for the morning.
Schedule brainstorming meetings in the afternoon.

Lion Daily Schedule

5:30 a.m.: Wake up, no snooze.

5:45 a.m.: Breakfast: high-protein, low-carb.

6:15 a.m. to 7:00 a.m.: Big-picture conceptualizing and organizing. Morning meditation.

7:00 a.m. to 7:30 a.m.: Sex. If you have kids who need help getting ready for school, make it a quickie.

7:30 a.m. to 9:00 a.m.: Cool shower, get dressed, interact with friends or family before heading to work.

9:00 a.m.: Small snack: 250 calories, 25 percent protein, 75 percent carbs. Ideally, have it at a breakfast meeting.

10:00 a.m. to 12:00 p.m.: Personal interactions, morning meetings, phone calls, emails, strategic problem solving.

12:00 p.m. to 1:00 p.m.: Balanced lunch. Go outside for sunlight exposure, if possible.

1:00 p.m. to 5:00 p.m.: Creative thinking time. Listen to music, catch up on reading and journaling. In a workplace setting, lead or attend brainstorming meetings.

5:00 to 6:00 p.m.: Exercise, preferably outdoors, followed by a cool shower.

6:00 p.m. to 7:00 p.m.: Dinner. Keep it balanced — equal parts protein, carbs, and healthy fats. A carb-heavy meal like pasta might make you crash.

7:30 p.m.: Last call for alcohol. A drink after this hour will knock you out.

7:00 p.m. to 10:00 p.m.: Socialize on the town, or connect with loved ones online while relaxing at home. You bought yourself an extra hour, so make the most of it!

10:00 p.m.: Be in your home environment by now. Turn off all screens to begin the downshift before bed.

10:30 p.m.: Go to sleep.

CHAPTER 4

A Perfect Day in the Life of a Bear

Ben,[1] 33, a married father of three from Los Angeles, was referred to me by his primary care physician. Twenty pounds overweight but otherwise healthy, Ben complained of low-grade but constant fatigue. As a supervisor at a home improvement big box store, he needed to be more alert than he was, and he hoped to have more energy for his kids after work and on the weekends. Even if he got a good night's sleep, he woke up groggy.

"I've got a physically demanding job," he told me at our initial consultation. "It's mentally demanding, too. I've got a lot to keep track of—deliveries, shipments, and paperwork. I need to be at the top of my game, but I never feel like I'm close to that. And then, when I get home at night, all I want to do is have dinner and relax. I want to get to the things I need to do around the house, play with my kids, but I can't get motivated."

Bears are social creatures. For their emotional health, they need to spend time with friends and family. I asked about Ben's friendships. "We used to hang out after work, but then we married off and started to have families. So now we usually hang out on the weekends. I'm in a baseball league on Saturdays, and that's always a good time. My wife and I have Saturday date night and go to dinner or a movie with other couples. Sundays are family day, and also my day of rest. If the kids don't jump on the bed and wake me up, I sleep in."

I asked Ben if he napped on weekends, too. "Oh, yeah! I fall asleep on the couch and wake up with potato chips stacked all over me. The kids think that's hilarious," he said. "Sunday nights are kind of rough, though. I can't fall asleep! I just lie there and think about the stuff I have to do on Monday."

Ben described the phenomenon called Sunday Night Insomnia. For a 2013 study, the online survey service Toluna Omnibus asked more than three thousand American adults, "What night do you have the most difficulty falling asleep?" Thirty-nine percent said Sundays. Most of them claimed to lie awake at least thirty minutes more on Sundays than on other nights. Saturdays came in a distant second, with 19 percent saying that it was the most difficult night for them to fall asleep.

Sunday Insomnia is a classic example of chrono-misalignment. By following a social schedule — like staying up late on Saturdays and sleeping in on Sundays — you throw your circadian rhythm out of whack, resulting in social jet lag and setting off a cascade of negative consequences from which it might take days to recover. By trying to catch up on sleep over the weekend, you wind up spending the entire week readjusting your circadian rhythm, ending with a net loss for the week's total sleep anyway.

Bears have a high sleep drive and need at least eight hours per night, or fifty-six hours per week, to ward off all the health risks of sleep deprivation, including weight gain, diabetes, heart disease, mood disorders, and low overall life satisfaction. If you sleep six hours a night for five days in a row, that'll get you to thirty. So on the weekends, you'd have to sleep for thirteen hours each day. That's not a viable plan, and you still won't get the sleep you're looking for!

Ben also falls into the weekend warrior trap of pushing off his fitness and socializing until Saturday and Sunday, another lifestyle choice that affects his sleep, metabolism, and energy level. "I always feel better if I can work out during the week, but it hardly ever happens," he said. "No time and no motivation. It seems really boring and kind of lonely to go to the gym. Not my style. I'd rather play sports with my friends or throw the ball around with my kids. And those things happen on the weekend."

Bears are creatures of the sun. Their chronobiology follows the solar cycle, meaning that when the sun comes up, their hormonal and cardiovascular systems respond, kicking their insulin, cortisol, and testosterone levels and their blood pressure and body temperature into high gear.

At their normal wake time of 7:00 a.m., they are fully functional and ready to begin their day. Then the sun goes down at six to nine o'clock, depending on the season, their bodies react to darkness again, and immediately, their endocrine and cardiovascular systems respond, starting the slowing-down process that culminates in sleep. Since Bears represent the majority, social schedules were built around their bio-time. Makes perfect sense. Half of the world's population gets hungry for dinner at 6:30 p.m., so it becomes the universal "dinnertime." Half of the world is ready for bed at 11:00 p.m., so it becomes "bedtime." Prime-time TV is between 8:00 and 10:00 p.m. because, during those hours, Bears are at low alertness and energy levels, and ready to plop in the couch.

If Wolves ruled the world, *Scandal* would air at 11:00 p.m.

If Lions were the majority, *Empire* would air at 7:00 p.m.

And if Dolphins dominated the planet? Well, we already have Netflix.

You'd think that by being in sync with the solar cycle and sticking with social schedules, Bears would already be operating at their peak potential. But it's a bit more complicated than that. Just because established social schedules exist does *not* mean that they match up with chronobiology. For example:

- Weekend warrior exercise habits are one of the reasons Americans are out of shape and overweight.
- Sleeping late on the weekends is a major cause of social jet lag and sleep deprivation.
- Eating the largest, heaviest meal of the day at 6:30 p.m. is one of the reasons Bears carry extra weight around their midsection.
- Scheduling business lunches and afternoon meetings at work guarantees subpar contribution from everyone attending, *especially* Bears.
- Sex at 11:00 p.m.? That's when your circadian rhythm wants you to be unconscious. I'm pretty sure most Bears would have a more satisfying sex life if they did it when their hormones and circulatory system weren't instructing them to pass out.

The framework of Bears' circadian rhythm matches social schedules, giving them a clear advantage over extreme Lions and Wolves. However, Bears are deeply susceptible to social jet lag. If they can make micro-adjustments to their schedules, they'll really feel the benefits.

The goals for Bears:

- **Get adequate sleep and exercise** *during the week.*
- **Shift eating rhythms to speed up metabolism and shed pounds.**
- **Increase energy in the afternoons and evenings with strategic napping and activity.**

If a Bear runs into trouble following the chronorhythm outlined below, it's due to:

- **Feeling hemmed in.** Bears might not like the idea of living by a clock. But the fact is, we all live by the clock in our brain, whether we want to or not. Instead of thinking you're a slave to a schedule, understand that real freedom comes from increased energy, dropping extra weight, better communication, and clearer focus. Freedom is about excelling and advancing in your life. If that means keeping track of when you eat, sleep, work out, talk, and think, it's a small price to pay for limitless possibility.
- **Taking long naps and sleeping in on the weekends.** If you just can't resist, I recommend sleeping in for an extra forty-five minutes on Saturday *only*, or taking a Sunday catnap of twenty minutes. Those small increases won't disrupt your chronorhythm. But wake up on Sunday at noon? Your entire week is blown.
- **Being tempted by late-night snacking.** Another habit that must be overcome for two key reasons: (1) It's the main cause of extra weight in the midsection, which itself raises your risk of diabetes, heart disease, and certain cancers. (2) Eating at night disrupts or delays sleep. Bears need their rest. If they don't get a solid eight hours, they can't function at full potential cognitively, creatively, and emotionally. A midnight snack can mess up your career and marriage? Yes, Bears. It can, by ruining sleep and making you fuzzy and irritable. This habit must be broken.

REALITY CHECK

The following schedule is how you'd organize your day in a perfect world. But real life is not perfect. Due to social or work situations that are out of your control, you might not be able to follow the schedule to the letter.

That's okay.

The worst thing you can do is say, "If I can't do **X, Y,** or **Z** at exactly the right time, the whole thing is thrown off, so forget it." *Any* changes will result in improvements to your health and happiness. It's not an all-or-nothing proposition. Ideally, you could do it all. Practically, you might not be able to. So do what you can now. Over time, as you notice positive changes, you might find that you can do a little more.

THE BEAR'S CHRONORHYTHM

7:00 A.M.

Typical: "The alarm goes off. I hit snooze a couple of times, and then I get up and start the day," said Ben.

Optimal: Wake up and have sex. In the early morning, your testosterone is high and your desire is strong. You might not be fully alert, but initiating sex upon waking is an excellent way to be active, elevate your heart rate, and raise core body temperature. Also, the boost of oxytocin in the morning will carry you through your whole day on a cloud of positive vibes, peace, and joy.

As an alternative to sex with or without a partner, elevate your heart rate immediately upon waking by putting on a pair of sweatpants and a t-shirt and walking around the block while you're still half asleep. If you wait to exercise until you're awake, you'll have the presence of mind to rationalize your way out of doing it. And you know exactly what I mean. If you can exercise outside, the exposure to sunlight will help you feel more alert. If you have kids who need help getting ready for school, do five minutes of sit-ups and push-ups on the bedroom floor. Every little bit helps.

7:30 A.M. TO 9:00 A.M.

Typical: "I do the morning routine. Shower. Have breakfast and two cups of coffee and drive to work in a fog."

Optimal: Have a healthy breakfast. It's a good idea to eat within a half hour of waking to sync the master clock in your brain to the minor clocks in your stomach and digestive system. Bears usually reach for high-carb choices like cereal or a bagel. Eating carbs in the morning raises calm-bringer serotonin and lowers cortisol levels, which you need to get up and moving. **Avoid carbs at breakfast.** Have a protein-heavy meal instead, like bacon and eggs, yogurt, or a protein shake. Don't be afraid to eat a hearty breakfast. People who eat the majority of their calories early in the day have a lower BMI than late eaters — even if they eat the exact same number of calories. To lose weight using the power of when, you will eat a big breakfast, a medium-sized lunch, a small afternoon snack, and a modest dinner, with zero late-night junk food binges. I'll make it easy for you. You'll see.

Also, avoid coffee at breakfast. I know it's a deeply entrenched habit. But coffee doesn't actually make you more awake first thing in the morning. It only makes you addicted to caffeine and jittery. Your commute will be safer if you get alertness from exercise, sunshine, and protein.

9:00 A.M. TO 10:00 A.M.

Typical: "I get to work and settle in. I make the rounds, talk to everyone, talk about what we watched on TV or what happened on the news. Not a lot gets done."

Optimal: Get your day organized. That foggy feeling of sleep inertia that paralyzes productivity when you first arrive at work? It's been eradicated with morning exercise or sex, sunshine, and a protein breakfast. Now you can actually plan your day in the first hour of work instead of wasting time.

10:00 A.M. TO 12:00 P.M.

Typical: "Finally, I start to feel fully awake. I'm already behind by then."

Optimal: Crank. Your cognitive peak comes mid-morning. Instead of frittering away mental sharpness socializing, tackle taxing work now to get it done in record time. If it's possible to close your office door or isolate yourself during this period, you can plow through paperwork. Have coffee now to further sharpen your alertness. One cup should be enough.

12:00 P.M. TO 1:00 P.M.

Typical: "My official lunch break. I love lunch. There are a bunch of options in walking distance, but I usually go right next door to a deli to get a sandwich."

Optimal: Exercise, eat, exercise. If you can move for thirty minutes before lunch by taking a walk, you'll speed up your metabolism to convert food to energy before you even take a bite, and decrease your appetite in one shot. Your meals should go in descending order by size, so your lunch should be half the size of breakfast and twice the size of dinner. If you're used to a twelve-inch sandwich at Subway, make it a six-incher. Ideally, you'll take another walk for ten minutes after eating, too.

1:00 P.M. TO 2:30 P.M.

Typical: "I feel okay at this time, energy-wise."

Optimal: Take charge. If you were active during your lunch hour, you can stave off the afternoon energy dip for an hour or two and prolong your on-peak analytical powers. Make good use of them until the inevitable lull hits.

2:30 P.M. TO 2:50 P.M.

Typical: "I always lose steam and feel really sleepy. But I need to be on my game, so I have a Coke or a Red Bull. Or a candy bar. Snickers is packed with energy, right?"

Optimal: Power nap. The best time to nap is approximately seven hours or so after waking. If you wake up at 7:00 a.m., the ideal nap time is 2:00 p.m. If you work at a progressive company like Google or the *Huffington Post* that has "sleep pods" for employees, or you have a home office, lie down and close your eyes for twenty minutes. The short nap will restore you to morning levels of energy and alertness. Make sure you set an alarm so you don't go longer than twenty minutes, or you'll wake up groggy with a second dose of sleep inertia, and you'll need another hour to shake it off. I know it's not possible for most of us to plan a nap, but if you can, your blood pressure will decrease and afternoon productivity will increase. At the very least, power down mentally for ten minutes. Find a quiet spot and do some deep breathing or meditation.

3:00 P.M. TO 6:00 P.M.

Typical: "I start watching the clock at around 3:00, and can't wait until I get to leave."

Optimal: Interact—and snack! If you make the recommended micro-adjustments, this will be the best time to attend meetings, interact with customers or clients, write emails, and make phone calls. If you were active midday and took a power nap, you'll be alert—able to concentrate on the needs and concerns of other people. Also, since it's toward the end of the day, a Bear-dominant workforce is already thinking about happy hour or dinnertime. If you have innovative ideas or strategies to present to the others, including the bosses, it's likely they'll be open to suggestions now. Take advantage of their agreeableness and get your thoughts and ideas approved.

At 4:00 p.m., have a small snack of around 250 calories that's 25 percent protein and 75 percent carbs (an apple with peanut butter, or cheese and crackers) for quick energy to last the final portion of the workday.

6:00 P.M. TO 7:00 P.M.

Typical: "Dinner! As soon as I get home, my stomach starts rumbling!"

Optimal: Exercise. At this hour, you're at your physical peak, able to access your maximum lung capacity and heart rate. Your hand-eye coordination is sharpest. As a friendly, sociable Bear, you can take advantage of your coordination high point by playing a team sport with friends. Join a post-work basketball league, take a class with a friend, or, if group sweating isn't your thing, use this Body Hour to run around and play with your kids or speed walk through errands.

Going in a completely different direction, this is also the best time for happy hour with friends. Your alcohol tolerance is high in the early evening, so you can lift a few glasses without getting too intoxicated. You'll also have time to metabolize the alcohol out of your system before it threatens to interfere with sleep.

7:30 P.M. TO 8:00 P.M.

Typical: "After a huge meal, all I really want to do is put on my sweatpants, sit down on the couch, and relax."

Optimal: Dinner and conversation. Dinner should be the smallest meal of your day, so have something filling, like soup or stew with a salad. Eating dinner one hour later than usual might seem like a challenge—you're hungry as a bear at 6:00!—but if you can delay the meal until 7:30 p.m., you'll be less likely to binge on junk food at 10:00 p.m. The main reason Bears tend to carry extra weight around their midsection is late-night visits to the fridge. If you can have the last bite of the night at or before 8:00 p.m., you'll speed up your metabolism, increase energy, and lose the belly fat. Eating within three hours of bedtime—late-night snacking—sends blood and heat to your core, which is a signal to the body to stay awake. The increase of digestion acids can cause heartburn when you lie down.

Also, if you don't overeat, and if you got exercise and sunlight several times during the day, as recommended, your good mood will continue. The day is more or less done, and now you are relaxed. Have potentially difficult conversations with family and friends now, when you—and other Bears—are catching a second wave of good vibes and positivity.

71

8:00 P.M. TO 10:00 P.M.

Typical: "On the weekends, my wife and I might be at a movie or a concert or have drinks with friends. But on weekdays, I'll most likely watch TV, play computer games, or go online until bedtime—and make many trips to the kitchen for snacks!"

Optimal: Brainstorm. When alertness and concentration are low, creativity is at its peak. Brilliant ideas tend to come into our heads when we're tired and groggy and doing just about anything besides sitting down and thinking really hard to try to come up with a brilliant idea. Your biological downtime is during the two hours before you get in bed. You don't have to do much to let the ideas filter in. One great place to brainstorm is in the bathtub. Not only will the soothing heat set your mind to wander, it'll reduce core body temperature, which will help you feel sleepy at bedtime. Other creativity boosters: reading, meditating, playing games, casual conversation.

10:00 P.M. TO 11:00 P.M.

Typical: "Still watching TV or online. Still snacking."

Optimal: Power-Down Hour. At 10:00 p.m., shut down all screens. Staring at blue wavelength light on phones and tablets at this hour suppresses melatonin secretions and will keep you awake. Instead, read a book, stretch, meditate. Have more sex.

11:00 P.M. TO 12:00 A.M.

Typical: "We get in bed at 11:00 and might watch the late news. My wife and I look at our Facebook feed and talk about the stuff our friends posted. If I have the energy, I might roll toward her and start something."

Optimal: Enter Phase One. Bears have a high sleep drive and need roughly eight solid hours a night. A belly full of snack food and the blue wave light influx could make it harder for you to fall to sleep. Any sleep-onset delay takes a toll. But, since you've already had morning sex and have powered down all screens, the only thing to do now in bed is go to sleep. Thanks to being active throughout the day, and not having a late-night bowl of chips and salsa, you should be able to fall asleep quickly and deeply. The first portion of the night is when your body is physically restored. Your cells heal, recover, and rebuild.

1:00 A.M. TO 3:00 A.M.

Typical: "I'm usually asleep by midnight, except on Sundays, I might not pass out until 2:00."

Optimal: Enter Phase Two. The second portion of the night is uncomplicated rest.

4:00 A.M. TO 7:00 A.M.

Typical: "Snoring, or that's what my wife says."

Optimal: Enter Phase Three. The third portion of the night is when you get the bulk of REM sleep. Muscles are inactive, making your throat narrower, which causes snoring. The excess weight you might carry is not helping in that department, either. While snoring, you're consolidating memory and clearing the cobwebs. If you follow the chronorhythm, you'll get a good three hours of the mental restoration you require to wake up refreshed and energized.

EASY DOES IT

It seems like a lot of change, and it is. But if you adjust your schedule slowly, making one or two small adjustments per week, you'll be able to incorporate and internalize them into your life seamlessly. You'll notice significant improvements in the overall quality of your life in just one month.

Week One

Set a consistent wake time and sleep time.
Shift your biggest meal from dinner to breakfast.
Go to www.thepowerofwhen.com to watch a video about waking up gently.

Week Two

Continue with previous week's changes.

Use early morning hours for practical work and the late evenings for creative brainstorming.

Socialize with colleagues in the afternoon instead of in the morning.

Week Three

Continue with previous weeks' changes.

Try to be active before and after each meal, even if it's just taking five-minute walks.

Don't eat or drink alcohol after 8:00 p.m.

Even if you party late on the weekends, continue to wake up within forty-five minutes of your regular time.

Week Four

Continue with previous weeks' changes.

Have sex in the mornings instead of late at night.

Take a twenty-minute power nap at 2:30 p.m.

Bear Daily Schedule

7:00 a.m.: Wake up, no snooze.

7:00 a.m. to 7:30 a.m.: Sex or exercise to elevate the heart and get cortisol flowing. Do it outside preferably (the exercise, or the sex if you're daring). If you don't have time for a twenty-five minute workout, five minutes is better than none.

7:30 a.m.: Breakfast, high protein, low carb. No coffee yet!

8:00 a.m. to 9:00 a.m.: Go to work. You'll have a safer commute if you replace caffeine with exercise in the a.m. If you work at home, get right to it.

9:00 a.m. to 10:00 a.m.: Plan and organize your day.

10:00 a.m. to noon: Most productive period. Concentrate, be on task, and get things done. Coffee break.

Noon to 12:30 p.m.: Non-exercise activity — walking is ideal.

12:30 p.m.: A medium-sized lunch. It should be half the size of breakfast and twice the size of dinner. Take a ten-minute walk after eating.

1:00 p.m. to 2:30 p.m.: One more hour of alertness before the afternoon energy dip.

2:30 p.m. to 2:50 p.m.: Power nap. If that's not possible, find a quiet place to do deep breathing exercises for a few minutes.

3:00 p.m. to 6:00 p.m.: Peak mood. Use your positive attitude in meetings, make phone calls, and send emails.

4:00 p.m.: Small snack, 250 calories, 25 percent protein and 75 percent carbs.

6:00 p.m. to 7:00 p.m.: Exercise if you didn't this morning, or be informally active by playing with kids or running errands. Or have drinks with friends.

7:30 p.m.: Dinner! A small meal that's filling, like soup or a bowl of stew with a salad.

8:00 p.m. to 10:00 p.m.: Socialize (sober; don't drink after 8:00 if you want high-quality sleep). Have light conversations. Take a soothing hot bath and let your thoughts drift. Maybe a brilliant idea will float into your head.

10:00 p.m.: Turn off all screens. Meditate, stretch, relax.

11:00 p.m.: Go to bed.

CHAPTER 5

A Perfect Day in the Life of a Wolf

I knew from the minute Ann[1] walked in the door that she was a Wolf. A forty-year-old mother of two, she was wide-awake at her 5:00 p.m. appointment, a time when most people are dragging. Her brain ran a million miles a minute, a hallmark of Wolves. They are quick thinkers and see every situation from multiple vantage points. Ann was carrying about thirty extra pounds, another Wolfish tip-off. Of course, not all Wolves are overweight. But due to their late-night eating tendencies and difficulties resisting temptation, they have a higher BMI than the other chronotypes do.

Her complaint was insomnia. "I get in bed at midnight and lie there for hours, my brain going crazy thinking about everything I have to do the next day and other stupid, random stuff," she said. "I finally fall asleep at two a.m. When the alarm goes off at seven, it's such a shock, I feel like I'm going to have a heart attack. I call it a 'bumpy reentry into reality.'"

Incidentally, most heart attacks and strokes happen between 4:00 a.m. and noon. Time of day is a critical factor in many health problems, including asthma attacks, arthritis flare-ups, epileptic seizures, heartburn, fevers, and more.

Ann forces herself out of bed to wake her daughters and her husband. After showering and getting herself dressed, she helps her kids get organized and makes breakfast for her family. "I'm in a complete fog," she said. "I'm doing it all on automatic pilot. I can pour cereal, but if someone asked me a hard question that required two brain cells to rub together, I'm lost."

Wolf

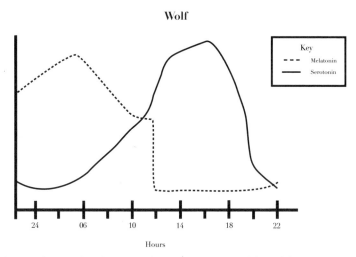

Key
- - - Melatonin
—— Serotonin

Hours

For Wolves, melatonin levels start to drop at 7:00 a.m. and don't fall off completely until noon. Serotonin levels peak in the evening, putting Wolves in a good mood at the end of the workday.

"I don't eat breakfast," she admitted when I asked about it. "No appetite. It's the only time of day I absolutely can't eat. I drink two cups of coffee while my family eats, and another on the road." Ann commutes by car to her job as a graphic designer at a small advertising firm in Scottsdale, Arizona. "I'm basically half asleep on the road," she said. "The only thing that keeps me from getting in an accident is caffeine."

Grouchy and foggy, Ann describes her morning work hours as "a waste. My physical body is there, and I can sit in front of the computer and do some stuff, but nothing approaching quality work gets done for hours," she said. "I can't pass for coherent until eleven-a.m.-ish."

In the late afternoon, Ann hits her stride. "Tea time, like four p.m., is my finest hour in the workday," she said. "My functional workday is only two hours long. I'm good, and I get enough done. But when I think about what I could do if I were inspired and alert in the morning, I'd be the head of the department, although I'm not sure I'd want to be.

"I leave the office around six and drive home at rush hour. This is when I start to feel wide-awake," she continued. "When I get home, I relieve the babysitter and try to talk to the kids about their day, but I can't concentrate, partly because my mind is still at work. I get a lot of ideas about what I could have done and need to do tomorrow when I should be listening to my kids, which makes me feel guilty and neglectful as a parent. My husband is exhausted after work. When I try to talk to him about important stuff, like bills and the kids, he always says, 'In the morning.' When he goes to bed, I stay up answering emails from friends and vowing to myself to get up early and hit the gym I've been paying for for five years but have been to exactly three times. I do open a bottle at dinner. After a few glasses, I'm good for nothing but snacking on the kids' junk food (one more handful), scrolling through Facebook (one more link), and binge-watching TV (one more episode)."

Ann complained about her inability to get things done and self-diagnosed as an insomniac. She had been to doctors (not sleep specialists) and had been put on different medications. Sleeping pills, antidepressants, even an antipsychotic. My diagnosis was not insomnia. Her ability to sleep wasn't broken. She could fall asleep, and when she was out, she reported a good quality of sleep, just not enough of it. **Her problem is being a Wolf in a Bear's world.**

Although Ann couldn't alter her daily work schedule too much, she could make a few simple changes to dramatically improve her performance at work, the quality of her relationships, her fitness, her outlook, and her overall health.

The goals for Wolves:

- **Improve efficiency during work hours.**
- **Shift eating rhythms to speed up metabolism.**
- **Increase number of hours of sleep per night.**
- **Stabilize mood swings for improved overall life satisfaction.**

Wolves are creative and completely open to trying new things. Following the chronorhythm I've created is, for them, a cool scientific experiment. But Wolves do run into some trouble sticking with it, due to:

- **Rebelliousness.** I would suggest to Wolves that they think of their chronorhythm not as a set of rules to rebel against but as a listing of absolutes. Gravity isn't a rule. It's an absolute. What happens when you rebel against gravity? You fall on your face. Your "when" is an absolute. If you stick with it, you'll soar.

- **Impatience.** Wolves can be emotionally sensitive, and if things don't go well, they might react negatively and blame themselves. That's the last thing I want to happen. The objective is to organize your life for greater happiness, not set yourself up for anxiety and depression. Be aware that change will come after a week's adjustment. Please be patient and allow the positive changes enough time to occur, and then reap the benefits.

- **Impulsivity.** Wolves have a hard time resisting temptation, and they make spontaneous decisions. Keep that adventurous spirit within the confines of a schedule. For example, one recommendation is to take a walk before and after meals. Be spontaneous about the direction you take and what food you choose, just as long as you take that walk and eat meals on your bio-time.

REALITY CHECK

The following schedule is how you'd organize your day in a perfect world. But real life is not perfect. Due to social or work situations that are out of your control, you might not be able to follow the schedule to the letter.

That's okay.

The worst thing you can do is say, "If I can't do **X**, **Y**, or **Z** at exactly the right time, the whole thing is thrown off, so forget it." *Any* changes will result in improvements to your health and happiness. It's not an all-or-nothing proposition. Ideally, you could do it all. Practically, you might not be able to. So do what you can now. Over time, as you notice positive changes, you might find that you can do a little more.

THE WOLF'S CHRONORHYTHM

7:00 A.M. TO 7:30 A.M.

Typical: As Ann said, "I wake up to the alarm and hit the snooze button two or three times. It's like I'm still dreaming when I throw back the covers to start the day."

Optimal: Drift. Set two morning alarms. The first wakes you up. The second goes off twenty minutes later. During those twenty minutes, lie in semiconsciousness and ride the last waves of REM sleep while your mind consolidates and restores. In this half-dreaming state, you are firing on all cylinders creatively and might drift into a brilliant idea. When the second alarm goes off, quickly jot down or voice-memo anything that crossed your mind. Word to the wise: If you use voice memo, speak slowly. Another benefit of a twenty-minute drift: At 7:00 a.m., your body temperature has not risen sufficiently yet for movement (remember you're on a slightly different schedule than the rest of the world). Waiting in bed will give you time to heat up, making your morning easier to take.

You might be thinking: *I don't have time to drift.* Buy those twenty minutes by not showering in the morning (more on that later).

7:30 A.M. TO 8:30 A.M.

Typical: "I can't eat a bite in the morning. The thought makes my stomach turn."

Optimal: Eat breakfast. In post-fast starvation mode, your body needs energy. If you don't give it nutrition, it will turn to another source — your own muscles. First, drink twelve ounces of water to quick-start your metabolism, heating up your digestion and core, helping to wake you up. Then eat some protein. A hard-boiled egg, protein shake, or cup of yogurt is fast and easy. Multiple studies prove that a good breakfast prevents overeating later in the day.

Do not drink coffee! Your cortisol and insulin levels are already high and doing their job to get you up to speed. Caffeine only makes you jittery in the morning when your "wake-up" hormones are flowing. Want proof? Ann has three cups and is still groggy for hours, but she says she is afraid that if she cuts back, her fog will last longer and be worse. This might be true for a day or two. But then the fog will clear. Caffeine also suppresses appetite, and Wolves must learn to eat within an hour of waking.

8:30 A.M. TO 9:00 A.M.

Typical: "My commute is a blur. It's like I'm driving through brain fog."

Optimal: Move. Even if you drive or take a train to work, get some outdoor movement time in first. Five to fifteen minutes of direct sunlight in the morning signals to the brain that it's time to wake up and stops the production of melatonin, the cause of the fog. Exercise also heats you up and increases cortisol and adrenaline circulation. A time-wise trick to sneak in sun and movement in the morning is to park your car a few blocks away, walk to a subway stop that's one farther than your usual stop, walk to the end of the driveway and get the paper, take your dog for a walk, or simply walk to the end of your street and back. Make sure you are breathing deeply. It helps set things in motion.

9:00 A.M. TO 11:00 A.M.

Typical: "Still in a fog, but it's slowly lifting. I can't concentrate, so I just drink more coffee. I read blogs, return emails, chat with friends and coworkers."

Optimal: Consolidate. By mid-morning, melatonin secretions have stopped, and your heart rate and blood pressure finally surge. The sleepy feeling dissipates by 10:00 a.m. If you've eaten breakfast and avoided coffee, you should be able to be productive. Since you're still off-peak, use the time to consolidate and plan what you will do when you shift to on-peak to hit the ground running. Listen to your previously recorded voice memos and flesh out some ideas. It's the ideal time to gather your thoughts.

On weekends, when you have more flexibility, mid-morning is the ideal time for sex. Testosterone hits its daily peak and libido is abundant. Since I don't recommend sleeping in on the weekends, a good morning routine would be: Wake at 8:00, eat at 9:00, sex at 10:00, coffee at 11:00. I'm not saying this is possible or preferable every weekend morning. People without kids or commitments aren't going to want to jump out of bed, especially if they stayed up late the night before. But while sleeping until noon on Sunday might feel good, it will wreck your bio-time for days after. Make your choices with full understanding of the consequences.

11:00 A.M.

Typical: "Still trying to clear my head."

Optimal: Coffee break. Your morning cortisol release has run its course, so caffeine will do you some good now. Drink it black. No need to add sugar and cream or throw in some cookies or a doughnut. Carbs will slow you down with a spike in blood sugar and insulin. One cup should be plenty. If you usually have had four to six cups by now, be sure to see my video on caffeine fading at www.thepowerofwhen.com.

11:15 A.M. TO 1:00 P.M.

Typical: "So hungry! I'm first out the door to get something to eat at noon and have anything with melted cheese on it. Since I skipped breakfast, I get a cookie for dessert."

Optimal: List and lunch. A Wolf's mental alertness is on the rise. If you had lunch now, you'd interfere with that biochemical alertness. Take care of busywork tasks, the things that don't require too much concentration or insight but still need to get done. Hydrate to push your hunger back a little and be productive. If you must snack, make it pure protein, like a protein bar, mixed nuts, or Greek yogurt. Keep it small!

1:00 P.M.

Typical: "So full from eating too much and too fast."

Optimal: Lunch. Before you eat, take a short walk to stimulate your metabolism. Choose a meal that is one-third carbs, one-third protein, and one-third healthy fat (a salad with grilled chicken or shrimp, a sandwich with one piece of bread, a fajita bowl, sushi) to keep your energy level up.

A low-glycemic lunch sets the stage for your most productive part of the day, especially if you are in a creative field. You'll hit the ground running and will be able to get a lot done. If possible, eat with colleagues or friends. Your mind is active now, and you'll be articulate and witty.

2:00 P.M. TO 4:00 P.M.

Typical: "Now I feel tired again. A crash from the sugar at lunch? Nothing a fourth cup of coffee won't fix."

Optimal: Crank. Wolves' workday really starts now. For two hours after lunch, you can get things done, but you are still not at peak alertness yet. That won't happen for hours.

4:00 P.M.

Typical: "Now I'm hitting my stride."

Optimal: Snack. It's been three hours since you last ate, and it's four hours until dinnertime. Have a snack to tide you over, but be careful about portion size. Too much could spiral you into an insulin crash and ruin your afternoon productivity.

4:15 P.M. TO 6:00 P.M.

Typical: "By the time I'm finally sharp and hitting my stride, the day is almost over. Everyone else is killing time until they can leave, and I'm just getting started. I race through, like trying to do a full day's work in two hours."

Optimal: Interact. Your energy rhythm is flowing. Compared to Lion, Dolphin, and Bear colleagues and friends who are lagging, you are on fire at the end of the day and will outshine them in meetings and in one-on-one discussions. Now is the time for Wolves to present ideas to the boss and colleagues.

6:00 P.M. TO 7:00 P.M.

Typical: "I'm bouncing off the walls. Delayed effects of the pot of coffee? I rush home, throw dinner together, and eat with the kids."

Optimal: Exercise. Wolves get an evening energy surge. Your reaction time, muscle strength, flexibility, and heart and lung efficiency are at their maximum. Use them. Take a long walk. Go to the gym. Walk the dog or take the kids to the park.

I understand that this is the "traditional" dinner hour, but Wolves have to break that mind-set. It's not healthy for them to eat this early. If you have kids, dinner will have to be on a staggered schedule. Feed the kids first, focus on them, and wait to have your meal at an appropriate bio-time. If you don't have kids, exercise instead of eating. Activity is a natural appetite suppressant. After a few days of delaying dinner, your stomach will adjust and hunger won't be an issue.

7:00 P.M. TO 8:00 P.M.

Typical: "The kids do their thing and my husband hits the couch. But I'm ready for some fun. I try to rally friends to meet me for a drink or go to a movie."

Optimal: Bond. Unencumbered Wolves should cool down after a workout by meeting friends or having a pre-dinner drink or social hour. Wolves with kids can help them with homework and play games with them. Although concentration might be a little difficult while cortisol is flowing, you can use your energy advantage to show loved ones you care.

8:00 P.M. TO 9:00 P.M.

Typical: "Wine o'clock. It's something I look forward to and it does help me calm down when my mind starts racing at night."

Optimal: Have dinner. In the late evening, early nighttime, your senses — especially taste — are most acute. Having dinner on the late side will be more satisfying for you and will prevent late-night snacking that packs on pounds. Drink your wine before and during the meal, but stop when you finish to give your body time to metabolize the alcohol before bed (drinking before bed can disrupt sleep). Or avoid wine and stay hydrated with water.

9:00 P.M. 11:00 P.M.

Typical: "Wine makes me hungry, so I'll have a snack or two while surfing the Internet or chatting online. I'm not reaching for cut veggies and fruit. More like the junk food or leftovers."

Optimal: Have fun (including sex). You'll be in the best mood all day, making it the perfect hour for relaxed or practical conversations with family and friends. Your body temperature is peaking, making you responsive to sexual activity. Sex brings physiological benefits to nearly every system in your body and releases hormones that bond you with your partner and make you feel happier for hours. Because you'll probably stay awake for a while afterward, you will benefit from these hormones instead of squandering them while asleep. The aerobic activity will quash hunger, making you less likely to snack late at night. Post-sex, do some household business. You'll have patience for work queries, annoying social correspondence, and balancing the budget.

11:00 P.M. TO 12:00 A.M.

Typical: "I'm probably online still, watching a show or reading articles, snacking. I start to worry that I should get to bed soon, but I feel wide-awake."

Optimal: Unplug. Using screens sends blue light into your eyeballs, suppressing melatonin release, keeping you awake. So finish up your emailing and shut down all screens. Spend the hour before bed meditating, reading, and stretching. In this meditative state, you will have your second creative peak of the day.

Showering or bathing at night not only buys you twenty minutes of drifting time in the morning, it will also help you fall asleep. Passive heating—taking a hot bath or shower—helps lower core body temperature, signaling to the brain to release melatonin, the key that starts the engine of sleep.

12:00 A.M.

Typical: "I lie in bed, listening to my husband sleeping. It's stressful not to be able to fall asleep, and I worry about what tomorrow will be like."

Optimal: Go to bed. By adjusting your eating and showering, coffee and alcohol patterns, exercise and screen time, you'll be able to fall asleep by 12:30 a.m. or thereabouts. It might take a couple of weeks, but you'll get there.

12:30 A.M. TO 2:30 A.M.

Typical: "Staring at the ceiling."

Optimal: Enter Phase One, when physical restoration occurs. Your body heals and damaged cells are repaired.

2:30 A.M. TO 5:00 A.M.

Typical: "Most likely asleep, thanks to the Ambien I took an hour ago."

Optimal: Enter Phase Two. The middle portion of a night's sleep is when your body and brain have uncomplicated rest.

5:00 A.M. TO 7:00 A.M.

Typical: "Finally, deeply out right before I need to wake up."

Optimal: Enter Phase Three. During the final portion of the night, you'll get the bulk of REM sleep. The brain's restoration and consolidation of memory takes place now. If sleep is delayed at the beginning of the night, you won't get enough REM sleep at the end, limiting its creative benefits and the total hours of brain restoration and organization.

EASY DOES IT

It seems like a lot of change, and it is. But if you adjust your schedule slowly, making one or two small adjustments per week, you'll be able to incorporate and internalize them into your life seamlessly. You'll notice significant improvements in the overall quality of your life in just one month.

Week One

Eat breakfast.

Get five to fifteen minutes of direct sunlight within an hour of waking.

Wean yourself off morning coffee. Go to www.thepowerofwhen .com for an instructional video.

Week Two

Continue with the previous week's changes.

Delay coffee until 11:00 a.m. at the earliest.

Delay dinner until 8:00 p.m.

Week Three

Continue with the previous weeks' changes.

Shift shower time from early morning to late night.

Make good use of your morning time by planning for the day ahead.

Week Four

Continue the previous weeks' changes.

Exercise in the evening.

Unplug all electronics by 11:00 p.m.

Wolf Daily Schedule

7:00 a.m.: Wake up to first alarm. Drift until second alarm twenty minutes later. When you get up, quickly write down or voice-memo any ideas.

7:30 a.m. to 8:00 a.m.: Get dressed, morning routine time.

8:00 a.m.: Eat a high-protein breakfast. Grab ten minutes of direct sunlight. No coffee!

8:30 a.m. to 9:00 a.m.: Get out the door and into the sunlight. A short morning walk to the car or train will help you wake up.

9:00 a.m. to 11:00 a.m.: Use the morning to consolidate and get organized. Your peak hours are yet to come, so prepare yourself now for your productive hours later.

11:00 a.m.: Coffee break, no snack. Carbs will only slow you down.

11:15 a.m. to 1:00 p.m.: Knock off all busywork tasks that don't require too much concentration or insight.

1:00 p.m.: Balanced lunch. Your brain and power of speech are sharp now. At lunch with colleagues, you'll be impressive and charming.

2:00 p.m. to 4:00 p.m.: Tackle hard tasks that require concentration.

4:00 p.m.: Snack of 250 calories, 25 percent protein, 75 percent carbs.

4:15 p.m. to 6:00 p.m.: Connect and interact with others. While their energy is waning, you're wide-awake and alert. Take advantage and attend or call meetings, make phone calls, and send emails.

6:00 p.m. to 7:00 p.m.: Exercise while your body is all warmed up for performance and safety from injury.

7:00 p.m. to 8:00 p.m.: Post-workout happy hour, pre-dinner social hour with friends. Homework hour with kids. You are up for anything now, so go do it.

8:00 p.m. to 9:00 p.m.: Dinner. By delaying the meal until now, you'll head off bingeing later on. Carbs will help calm you down for sleep.

9:00 p.m. to 11:00 p.m.: Best mood of the day, and time for fun, including sex.

11:00 p.m.: Turn off all screens. Relax, meditate, read, stretch, take a hot shower or bath.

Midnight: Go to bed.

THE BEST TIME TO DO EVERYTHING

This section is a guide for tapping into the power of when to perform at your best in several categories — Relationships, Fitness, Health, Sleep, Eat and Drink, Work, Creativity, Money, and Fun. Within each chapter, you'll find specific activities. Each activity stands alone, so you can feel free to go directly to what interests you or read straight through. However you choose to take in the information, it'll make you rethink the timing of everything in your life.

CHAPTER 6

Relationships

FALL IN LOVE

Failure: Not being able to find, affirm, and maintain romantic intimacy.

Success: Seeking, solidifying, and sustaining romantic intimacy.

The Simple Science

Sorry to be unromantic, but the truth is, falling is love is a biochemical process. Let's take a look:

The attraction rhythm is when you feel the flush of new love. It can happen at any time. In fact, it's right under your nose. Pheromones are odorless hormones that both men and women emit. We can't detect them the way we can inhale the fragrance of a rose or an orange, but pheromones come in through our noses and go straight to our brains. Call it "love at first smell." If a particular person responds to another's pheromone signature, he or she feels sexual attraction. You can fall in love with someone's looks from across the room, but until you're standing within nose range, you can't know if there's any "chemistry" (pheromone attraction) between you.

That attraction might be affected by bio-timing. According to a University of Texas at Austin study,[1] men can detect a woman's fertility based on smell alone. Researchers asked premenopausal women to sleep in t-shirts for three consecutive nights during ovulation, and in a different tee for three nights during a non-ovulatory span. Male subjects were then asked to smell the t-shirts. They described the ones worn by women who were ovulating as more "pleasant" and "sexy" than the others. So, women, an important attraction rhythm for you is not to use perfume or scented bath products during ovulation.

Not to suggest looks don't matter at all. They do, just not in the way you might think. The most sexually attractive facial expression is kindness. In a 2014 Chinese study,[2] researchers showed pictures of faces to 120 participants (half men and half women) and asked them to rate the images on attractiveness. The photos of kind and positive-looking people

were rated as more attractive than those that didn't convey these qualities, leading researchers to call the goodness-is-beautiful phenomenon
the halo effect.

What does this mean in terms of attractiveness and bio-time? **Search
for new partners when you're in a good mood.**

- **Dolphins** are in a good mood in the afternoon and into the
evening.
- **Lions** are in a good mood in the morning and early afternoon.
(Lions' worst moods are often better than other types' best moods. So a
bad mood for them isn't really that horrible in comparison.)
- **Bears** are in a good mood starting in mid-afternoon and continuing
into early evening.
- **Wolves** are in a good mood starting in late afternoon and
continuing into late evening.

You can see how the dinner date became standard. Most chronotypes
are in a good mood later in the day.

The affection rhythm is when you feel affectionate and are motivated
to touch and be close with a partner. It involves several hormones, including
dopamine, serotonin, vasopressin, and, most significantly, oxytocin. During the early stages of a relationship, when couples cuddle, hold hands,
hug, and kiss constantly, the oxytocin flows like a river. According to a
2012 Israeli study,[3] researchers measured the oxytocin levels in sixty
new couples three months into their burgeoning relationships compared
to forty-three singles. The couples' oxytocin levels were significantly
higher than the singles'. The rise in oxytocin levels correlated with the
happy pairs' positive affect (general happiness) as well as their anxiety
about the relationship. The gist being, when you love a lot, you also
worry about your partner and where the relationship is going. Speaking
of which...

The attachment rhythm is when you bask in a long-term bond with
your partner, and it can be measured in blood, too. The Israeli researchers
retested the couples who were still together six months later (roughly half

of them) and found that their oxytocin levels had not decreased. The lead researcher, Ruth Feldman, told *Scientific American*, "Oxytocin can elicit loving behaviors, but giving and receiving these behaviors also promotes the release of oxytocin and leads to more of these behaviors." The attachment rhythm is a positive feedback loop that starts with attraction and is reinforced daily with affection. Whenever possible, show daily physical affection and facial kindness to your partner to keep the love alive.

The Rhythm Recap

The attraction rhythm: When you feel drawn to a new love interest, thanks to pheromones and the halo effect.

The affection rhythm: When you feel the urge to be physically close with a new partner, thanks to the spike in love hormones.

The attachment rhythm: When you feel a long-term bond with a partner, thanks to the steady flow of love hormones.

THE WORST TIME TO FALL IN LOVE

11:00 a.m. to 2:00 p.m. Morning oxytocin, testosterone, and dopamine levels have receded by then (even for Wolves). For all chronotypes, positive affect is lower at the heart of the workday. Lunch dates may be a great way to meet a person and get to know him or her, but dinner is when you become attracted.

THE BEST TIME TO FALL IN LOVE

Dolphin: 8:00 p.m., in the afterglow of a serotonin-boosting dinner of mainly carbs, and after sex, which releases oxytocin.

Lion: 7:00 a.m., after morning sex.

Bear: 4:00 p.m., in the afterglow of a nap, brimming with positive affect.

Wolf: 11:00 p.m., in the afterglow of a serotonin-boosting dinner and oxytocin-producing sex.

CALL A FRIEND

Failure: Calls going straight to voicemail, not getting calls back, and/or reaching friends when they're in no mood or have no time to talk.

Success: Reaching out to friends when they're available and ready to chat.

The Simple Science

Calling a friend seems like it should be straightforward. You dial the number, or tell Siri to do it for you. Your friend answers — since the phone is already in her hand or nearby — and the two of you have a lovely conversation. You make plans or have a satisfying exchange of thoughts, news, and ideas, and then hang up feeling loved and appreciated.

It doesn't always work out that way. How many times have you called someone, and, instead of a friendly voice, you get an automatic reply like, "I can't talk now. I'll call you later?" Knowing or suspecting you've been screened can leave you feeling rejected or neglected.

As social creatures, we need to make social connections. We're hard-wired to seek out and link our lives to others for survival. Nowadays, we don't band together to hunt for food. But we need adequately maintained friendships to ward off loneliness, which can lead to depression, immune system disorders, and disease-causing inflammation.[4]

Texting, the preferred method of connection by young adults, is a poor substitute for face-to-face interaction. Voice-to-voice is the next best thing, with its richness of tone and nuance. Nothing is as friendship-affirming as hearing the sound of you and a pal laughing together.

So, when to make that call to guarantee a hit of positive interaction? The first factor to consider is **the availability rhythm.** When are your friends free to chat? According to 2013 research from the University of Southern California's Marshall School of Business,[5] the vast majority of the study's 554 professionals consider answering a cellphone during a formal or informal meeting to be inappropriate. Checking incoming calls, excusing yourself to take a call, and even bringing a phone into a meeting were all considered disrespectful and disruptive. Millennials were three times more likely to think it was okay to check their phones or take calls at work than those over forty (that is, their bosses), a stat that could have far-reaching consequences for new adults' careers. **If you want to be a good friend, don't put anyone in an awkward situation by calling during work hours.**

Don't call a new mother when she or the baby might be napping. Don't call a student when she is likely studying for an exam the next day. Don't call between 6:00 and 9:00 p.m., when most of us are having dinner. And don't call during your friend's favorite shows or on game day. That's just annoying.

There's also an **intimacy rhythm** to calling people. In a 2015 Cornell University study,[6] researchers examined not only when calls were placed but to whom, and concluded that the later the call, the more intimate the relationship. Late-night calls were placed only to the subjects' closest friends and romantic interests. In fact, calling late is a way to subconsciously telegraph that you consider the relationship special. If you call someone during daylight, the unconscious perception is that you have an agenda, such as to make a plan or exchange necessary information. If you call after sundown, the unspoken understanding is that you want to share stories and have a deeper, more intimate dialog. If you call a friend to shoot the breeze at 2:00 p.m., it might come across, even subconsciously, as odd or intrusive.

Last, try to use the **bio-time calling rhythm,** which involves adjusting your calling patterns to those of different chronotypes. Calling a Lion at 10:00 p.m. would be the equivalent of calling a Wolf at 2:00 a.m. or a Bear at midnight. You'd have to be *very close* with a friend to expect him to chat with you when he's half asleep. Same thing with calling too early. You Lions might be up at 6:00 a.m. and eager to catch up. But your Bear and Wolf friends? No. The Cornell study also looked at chronotype phone-calling preferences and confirmed that **morning types are motivated to call friends and engage socially during the day, while evening types do so at night.** If your aim is to reach someone when he or she is most likely to chat, think of his or her chronotype before your own. Obviously, your chronorhythm matters as well, so think about what you are trying to accomplish with your call. If you want to have a fun, light-hearted conversation, call when you're slightly tired or later at night. For important conversations that require focus, call at lunchtime or mid-afternoon, when all chronotypes are more alert.

The Rhythm Recap

The availability rhythm: When someone is most likely to have the time and attention to answer a call, based on work and family schedule.

The intimacy rhythm: When to call someone, based on the depth and closeness of the relationship.

The bio-time calling rhythm: When to reach out, according to the chronotype of the person receiving the call.

THE WORST TIME TO CALL A FRIEND

9:00 a.m. to 3:00 p.m.; 6:00 p.m. to 9:00 p.m. Don't make social calls during work and school hours. If you must contact a friend during the workday, text. Also, don't call at dinnertime, when someone might be spending quality time with family or might be out on a date.

THE BEST TIME TO CALL A FRIEND

Since your goal is to place phone calls that strengthen social relationships, the question is, "When will your friend be most happy to hear from you?" Don't call according to *your* chronotype. Call according to his or hers. If your friend is a . . .

Dolphin: 9:00 p.m. to 10:00 p.m.

Lion: 7:00 a.m. to 10:00 a.m.

Bear: 8:00 p.m. to 10:00 p.m.

Wolf: 9:00 p.m. to 11:00 p.m.

THE BEST TIME TO CALL PARENTS AND GRANDPARENTS

For all chronotypes, the answer is: "Anytime."

As a father and a psychologist, I advise you to call senior relatives whenever it's convenient for you during their waking hours. Don't worry about availability. Focus on frequency.

Socially isolated seniors are among the most vulnerable population in society. Setting aside the obvious, such as cognitive deterioration and death, they're at higher risk than others for cardiovascular disease and infectious diseases. Senior loneliness increases the risk of heart disease, elevated blood pressure and cortisol level, a heightened response to stress, and depression. A 2012 University College of London study[7] followed 6,500 adults fifty-two and older over a seven-year period and found that isolated and lonely subjects were almost twice as likely to die as those who received regular calls and visits from family and friends. Maybe it's because the old and alone don't take good care of themselves, or maybe psychological pain takes a physical toll. Probably both. Not to guilt you into it, but calling older relatives and asking them how they're doing can save lives.

According to the Cornell University mobile phone use study mentioned above, people are less likely to call family in the evening, when they prefer to talk to their closest friends. The theory is that family relationships aren't as fragile as friendships. You can call your parents and grandparents pretty much whenever you feel like it, which winds up being during daylight hours. That works for seniors. After age sixty-five, most of us shift into a Lion or Dolphin bio-time (more on that in "Chrono-Longevity," page 327) and tend to wake up early, eat early (blue plate special, anyone?), and turn in early. Might as well check in with seniors early, too.

FIGHT WITH YOUR PARTNER

Failure: Destructive arguments that don't resolve conflicts but do fill you with bitterness and resentment.

Success: Constructive arguments that resolve conflicts and cement your bond.

The Simple Science

Couples or close friend pairs don't always agree, and why should they? Any two different people will have differences of opinion. Arguing with your partner is an unavoidable part of life. You can do it the healthy way, with open, honest, fair communication and a commitment to conflict resolution and compromise. Or you can yell and scream and hurl accusations and sarcastic barbs. The time of day you pick fights will influence whether a heated discussion makes your relationship stronger or breaks your union apart.

The first and most important time-wise tip: Don't pick a fight unless you've had a decent night's sleep. Call it **the sleep-deprivation rhythm** — the "you're overreacting" rhythm. In a Tel Aviv University study,[8] researchers recorded sleep-deprived subjects' brains with both MRI and EEG machines while the subjects did emotional-cognitive tasks. The results showed that, when sleep-deprived, subjects were unable to control their emotional reactivity. Things that should have registered in their minds as "neutral" instead registered as "negative." Nothing good will come of an argument between two exhausted people. It's better to shelve it completely until you're better rested.

What about **the resolution rhythm** — the "don't go to bed angry" rhythm? That old saying should be changed to "don't argue before bed." During deep sleep, the brain consolidates memory. If you have a big screaming fight at midnight and then go to sleep, your brain will cement the negative emotions of the fight in your mind overnight *even if you made up beforehand.* In a University of Massachusetts study,[9] the responses of

106 men and women were recorded after the subjects were shown emotionally provocative images—from disturbing to positive—and then the subjects' memories and emotional responses were tested when they were shown the same images a second time—after either a good night's sleep or a full day of wakefulness. The negative emotional responses to the disturbing images were stronger in the subjects who'd slept the night than in the ones who'd stayed up. Losing a night's sleep due to fighting might be better than fighting, resolving the issue, and then falling into a deep sleep. Don't take this the wrong way. I'm not advising you to fight all night long. Ideally, you'd argue (or talk productively) earlier in the evening and resolve the conflict three hours before sleep, giving you adequate time to have other experiences and memories before sleeping.

The mood rhythm, or "why are you doing this now?" rhythm, can predict the outcome of an argument before you fire the first shot. We all understand that mood, in a nonclinical sense—how you're feeling at any given moment—is fluid. Moods change over the course of the day based on what happens but also based on bio-time. According to a study done at the University of Warsaw, Poland, in 2008,[10] chrono-mood can be accurately measured using a three-dimensional model. The first dimension is "energetic arousal," or how energetic or tired you are. The second is "tense arousal," or how nervous or relaxed you are. The third is "hedonic tone," or how pleasant or unpleasant a mood you're in. The Polish researchers looked at the diurnal (daylight) shifts of the three mood markers from 8:00 a.m. to 8:00 p.m. in nearly five hundred adult men.

The morning preference subjects' energy level started high, hit its peak midday, and then turned sharply downward until evening. They were more relaxed throughout the day than evening types, and their tension peaked in the evening. Their pleasantness started higher than that of the evening types, too, and shifted only slightly as the day wore on, with their best mood hitting around 10:00 a.m. and worst mood (although it was well above the evening types') at 6:00 p.m. In sum, they were tired, tense, and cranky in the late afternoon into mid-evening. **Do not rattle a Lion's cage between 3:00 p.m. and 8:00 p.m.**

The evening types' energy started low and rose steadily throughout

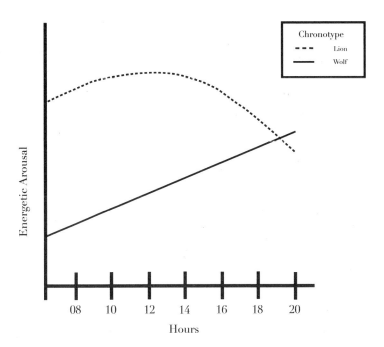

Lions' energy levels start high and then decline gradually from midday until bedtime. Wolves' energy levels start low and don't peak until well after Lions have already gone to bed.

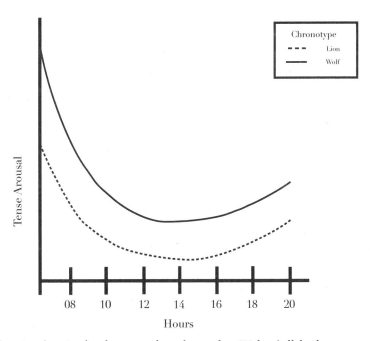

Lions' anxiety/tension levels start and stay lower than Wolves' all day long.

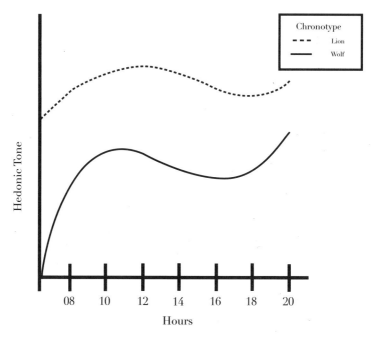

Lions' hedonic tone—how pleasant their mood is—starts and stays higher than Wolves' all day long. Wolves' moods have greater fluctuation than emotionally stable Lions'.

the day, peaking in the evening. They started at a higher level of tension than morning types, with peak relaxation hitting at 4:00 p.m. and swinging upward until evening. Their pleasantness level swung throughout the day, with a low at 8:00 a.m. and another dip at 6:00 p.m., and with peaks at noon and 8:00 p.m. In sum, Wolves were tired, tense, and cranky throughout the entire day but at their worst in the morning. **Call out a Wolf before noon at your peril.**

For their part, Dolphins and Bears are conflict-averse and tend to avoid fights as much as possible. They will seem frustratingly distant and evasive during sleep inertia, which can last deep into the morning. Same thing for Bears during their post-lunch afternoon energy dip around 2:00 p.m. **Do not pick a fight with a Dolphin or a Bear until after 4:00 p.m., or it'll seem like you're fighting with yourself.**

The self-regulation rhythm, or "I can't stop myself," is about the behavioral capacity to think long-term and put aside the emotion of the moment, or to just say whatever comes to mind without thinking of future ramifications: "I have no filter." Lions, Dolphins, and Bears rate lower on impulsivity and future-orientation[11] than Wolves, but they, too, are vulnerable to saying things when they're off-peak that they can't take back. Wolves, on the other hand, are unlikely to hold their tongues ever and will say anything to win an argument *now*. When Wolves are off-peak (mornings and mid-afternoon), they might be crankier, but they'll also be slightly less acidly articulate. When they're on-peak and looking for a fight, hide.

The Rhythm Recap

The sleep-deprivation rhythm: When your emotions are muddled due to lack of adequate rest.

The resolution rhythm: When to resolve a fight so it doesn't linger in your mind.

The mood rhythm: When mood affects the likelihood of getting in an argument, and its intensity.

The self-regulation rhythm: When chronotypes can and can't hold back from saying something they'll regret.

THE WORST TIME TO FIGHT WITH YOUR PARTNER

11:00 p.m. Fighting before sleep, when you're both tired and therefore overly sensitive and, for Lions and Bears, at mood low points, is not a good idea. Even if you make up before you turn off the light, sleep immediately after a fight will cement the negative emotions in your mind. Better to argue constructively in the late morning or early afternoon.

THE BEST TIME TO FIGHT WITH YOUR PARTNER

Dolphin: 7:00 p.m. Dolphins are conflict-averse and good listeners. Wait until after a carb-heavy dinner and they'll agree to just about anything.

Lion: 9:00 a.m. Lions will be alert, analytical, and eager to fix things, although they might not listen well.

Bear: 5:00 p.m. Bears might not understand what the fight is about, but they'll be most willing to compromise when in a good mood.

Wolf: 8:00 p.m. Wolves are wide-awake and sharply articulate at this hour, but they're also in their best mood of the day. Proceed with caution.

Fight with Your Partner Compatibility Chart

For productive discussion that leads to positive resolution:

You	Dolphin Partner	Lion Partner	Bear Partner	Wolf Partner
Dolphin	7:00 p.m.	7:00 p.m.	5:00 p.m.	7:00 p.m.
Lion	7:00 p.m.	9:00 a.m.	3:00 p.m.	5:00 p.m.
Bear	5:00 p.m.	3:00 p.m.	5:00 p.m.	5:00 p.m.
Wolf	7:00 p.m.	5:00 p.m.	5:00 p.m.	8:00 p.m.

"THE FRINGE BENEFIT OF LESS FIGHTING?"

"I laughed out loud when Dr. Breus told me that Wolves will say anything without thinking during a fight," said **Ann, the Wolf.** "That is so me. When I get going, crazy things come out of my mouth. I always feel bad about it later, and wish I could take it all back. But then, when my husband—a Bear, by the way—and I get into it again, I throw the kitchen sink at him. I'm at my meanest at night when I have the most energy, which I already knew. My husband is totally relaxed then, and the discrepancy is one of the things we fight about. I get mad at him when he doesn't have the energy to talk, or go out, or do things around the house that need to get done. And he gets mad at me for nagging him. As soon as he says 'nag,' it's like a siren goes off in my head. So we've decided that he has to respect my bio-time, and I have to respect his. I won't nag him to do things when he's relaxing, and he won't call me a nag, ever. We'll just do our thing on our own bio-time, and have our 'productive discussions' in the afternoons only. So far, so good! If I'm putting my energy toward writing at night (something I've always wanted to do), I don't get that angry at him anymore. The fringe benefit of less fighting? More sex."

HAVE SEX

Failure: Having an unsatisfying or nonexistent sex life.

Success: Having sex that is satisfying and frequent and provides you with enormous health benefits.

The Simple Science

How on earth did humans adopt the practice of having sex at bedtime, "where" and "when" we're supposed to be unconscious?

One argument is that sex is a sleep aid. There is not much science to back this up. As an expert in sleep medicine, I can attest to the fact that getting in bed at night and turning off the light does bring sleep. Having fifteen to thirty minutes of sex is beside the point. Melatonin goes up when the lights go off. If you have sex with the lamps blazing, sex can delay the onset of sleep for women. Making love while fighting sleep does not increase intimacy between partners. For Dolphins and Wolves, physical arousal at bedtime triggers wakefulness and can cause insomnia. You'll wind up staring at the ceiling and listening to your partner's snores.

In a recent study on sexuality, when subjects were asked why they had sex at a certain time,[12] **the convenience rhythm**—they were already in bed, their partner was available, and sex didn't interfere with their work schedule—accounted for 72 percent of sexual encounters. Only 28 percent of encounters happened because subjects felt sexual. And why would someone feel sexual between 11:00 p.m. and 1:00 a.m., when the vast majority of subjects had sex? That's when your heart rate is slow and melatonin is making you sleepy. Your body is not primed for *any* physical activity at this time, let alone sex. Turning your partner down night after night out of exhaustion or lack of desire can lead to hurt feelings and emotional distance. Going through the motions doesn't engender a loving feeling, either. Lackluster sex certainly doesn't strengthen your desire or make you want to have more of it.

Although we've been conditioned to associate sex with bedtime, **the**

desire rhythm peaks in the morning,[13] when testosterone in both men and women is at its highest (it's lowest at bedtime). Sexual fantasies mostly occur overnight and in the early morning, with the surge of testosterone, which is why most men wake up with erections. Sex upon waking will jump-start your day, fill you with energy, reduce stress and anxiety, and flood your brain with loving, happy hormones for hours. Not for nothing, creativity is enhanced now, too. You'll have all kinds of new, fun ideas for improved performance and greater satisfaction.

Good sex, when desire is peaking and you're physically and mentally alert, has enormous health and emotional benefits. A healthy **afterglow rhythm** increases circulation, oxygenates the entire body, and gives you a sense of well-being. Antibodies released during sex boost the immune system, preventing and curing minor ailments. Orgasm triggers oxytocin, elevating your mood and sense of connection to your partner all day. When oxytocin levels go up, cortisol levels go down. They're on a seesaw. More sex, less stress—and less of all the health problems associated with stress, such as obesity, heart disease, and mood disorders. The chemical benefit of sex, that loving feeling, is too often squandered if you have sex and then fall asleep. For Dolphins and Wolves, however, since they'll be awake until midnight or later, sex in the evening can promote relaxation and reduce stress.

What about **the masturbation rhythm?** You don't need a partner to have sex, after all. In fact, most American men (94 percent) and women (85 percent) masturbate,[14] whether they have a partner or not. For greater hormonal benefits, masturbate on bio-time or when desire is highest (6:00 a.m. for Lions, 8:00 a.m. for Bears, and 10:00 a.m. for Wolves) or when you need to reduce stress or boost your mood (early evening for Lions, early afternoon for Bears, early morning for Wolves). And Dolphins? To lower cortisol levels, masturbate an hour or two before bed, around 8:00 p.m.

The Rhythm Recap

The convenience rhythm: When couples have sex simply because they're both in bed and available.

The desire rhythm: When testosterone levels are at their highest, strengthening desire.

The afterglow rhythm: When oxytocin and other chemicals are flowing post-sex to give you a sense of well-being.

The masturbation rhythm: When to do it by yourself.

THE WORST TIME TO HAVE SEX

11:00 P.M. to 1:00 a.m., when 50 percent of sexual encounters happen.

THE BEST TIME TO HAVE SEX

Dolphin: 8:00 p.m.

Lion: 6:00 a.m. to 7:00 a.m.

Bear: 7:00 a.m. or 9:00 p.m.

Wolf: 10:00 a.m. or 10:30 p.m.

REALITY CHECK

The bio-timing of sex isn't always compatible with life. Unless you have a very secluded office space, having a quickie or masturbating at 10:00 a.m. is not advisable for Wolves. You might be so used to having sex immediately before sleep that the idea of morning sex is distasteful. Or, you might love the idea of morning sex but find you can't fit it into your schedule. That's fine. No worries. Just be aware that sex might be more satisfying and healthful at certain times. That doesn't mean it's not satisfying or healthful at other times. I'd never want anyone to have *less* sex simply because they feel compelled to stick to their chronorhythm. Spontaneity is just as important as testosterone, if not more so. If you're in the mood, go for it.

Sexual Chronotype Compatibility Charts

The following three charts—one for heterosexual couples, one for gay male couples, one for gay female couples—is based on a study[15] by researchers at the University of Warsaw of 565 subjects between the ages of eighteen and fifty-seven. The times in each box are **self-reported preference times for sex based on desire, not convenience.** The most important distinction is that women of all chronotypes feel a stronger need for sex between 6:00 p.m. and midnight, and Lions only report strong desire in the early mornings. Men of all chronotypes, on the other hand, feel a strong need for sex in the morning and evening. Even Wolf men were into it at 9:00 a.m. Male Lions, even exhausted, would have sex at midnight.

In the heterosexual box, I have evening as the number one choice, because if it's equally appealing to men to have sex in the morning or the nighttime, they might as well defer to the female preference of evening. Since men tend to prefer morning sex, the male gay couple chart includes morning times; the opposite applies for the female gay couple chart, with the one exception of female Lions.

Heterosexual Couple	Male Dolphin	Male Lion	Male Bear	Male Wolf
Female Dolphin	8:00 p.m./ 8:00 a.m.	8:00 p.m./ 7:00 a.m.	10:00 p.m./ 8:00 a.m.	8:00 p.m./ 9:00 a.m.
Female Lion	7:00 p.m./ 7:00 a.m.	6:00 p.m./ 6:00 a.m.	8:00 p.m./ 7:00 a.m.	7:00 p.m./ 8:00 a.m.
Female Bear	8:00 p.m./ 7:30 a.m.	9:00 p.m./ 7:30 a.m.	10:00 p.m./ 7:30 a.m.	10:30 p.m./ 8:00 a.m.
Female Wolf	9:00 p.m./ 9:00 a.m.	9:00 p.m./ 9:00 a.m.	10:00 p.m./ 9:00 a.m.	11:00 p.m.

Male Gay Couple	Dolphin	Lion	Bear	Wolf
Dolphin	8:00 a.m./ 8:00 p.m.	7:00 a.m./ 8:00 p.m.	8:00 a.m./ 10:00 p.m.	9:00 a.m./ 10:00 p.m.
Lion	7:00 a.m./ 8:00 p.m.	6:00 a.m./ 6:00 p.m.	7:00 a.m./ 9:00 p.m.	9:00 a.m./ 9:00 p.m.
Bear	8:00 a.m./ 10:00 p.m.	7:00 a.m./ 9:00 p.m.	7:30 a.m./ 10:00 p.m.	10:00 a.m./ 11:00 p.m.
Wolf	9:00 a.m./ 10:00 p.m.	9:00 a.m./ 9:00 p.m.	10:00 a.m./ 11:00 p.m.	11:00 a.m./ 11:00 p.m.

Female Gay Couple	Dolphin	Lion	Bear	Wolf
Dolphin	8:00 p.m.	8:00 p.m./ 8:00 a.m.	9:00 p.m.	10:00 p.m.
Lion	8:00 p.m./ 8:00 a.m.	6:00 p.m./ 6:00 a.m.	9:00 p.m./ 7:00 a.m.	9:00 p.m./ 9:00 a.m.
Bear	9:00 p.m.	9:00 p.m./ 7:00 a.m.	7:30 p.m.	10:00 p.m.
Wolf	10:00 p.m.	9:00 p.m./ 9:00 a.m.	10:00 p.m.	11:00 p.m.

PLAN SOMETHING IMPORTANT

Failure: Getting stuck in research, details, or the idea stage, or making a snap decision that you live to regret.

Success: Working as a team to conceive, research, and pull the trigger on a great plan you both love.

The Simple Science

Most of us don't see eye to eye about the division of labor and the details when planning a vacation, a wedding, or something vague like "the future." To successfully plan something important as part of a couple, a group of friends, or a family, **stick to two time-wise tricks:**

1. Assign specific jobs to specific chronotypes.
2. Execute the planning at the times of day that correspond to the specific task.

Regarding the first point, planning requires a wide range of skills—alertness, attention to task over time, flexibility, and creativity. Rarely will you find one person or one chronotype excelling in *all* of these skills. Each chronotype brings specific planning expertise to the table.

• **Dolphins** are perfectionists and obsessives. **Positives:** They excel at research and will exhaustively research the best hotels, flights, and rates. **Negatives:** They might feel overwhelmed by the options and will have trouble committing.

• **Lions** are executive decision makers and in-control types. **Positives:** They will pull the trigger on a plan and make all the arrangements. **Negatives:** They might plan a trip *too* carefully and leave little room for spontaneous exploration.

- **Bears** are happy with just about any decision that the Dolphins and Lions make. **Positives:** Their enthusiasm and flexibility are appreciated. **Negatives:** Enthusiasm and flexibility don't get the job done.

- **Wolves** are spontaneous and creative. **Positives:** They provide outside-the-box ideas and a spirit of adventure, regardless of cost and logistics. **Negatives:** Impulsively saying, "Let's just book it right now!" might turn out great, or it might lead to disaster.

Regarding my second point, about doing certain tasks at certain times of day, keep in mind:

The vision rhythm. When is the best time to envision your vacation, your wedding, or your future — to let your imagination run wild about where you'd like to go and what you'd like to do?

In a 2011 study, researchers at Albion College[16] asked subjects to answer certain types of questions at different times of day and lined up their ability to answer correctly by their morningness or eveningness preference. The problems they were asked to solve were either analytical (logic, math) or insightful (aha! realizations).

Here's an analytical question that will bring you back to middle school math (sorry): Train A leaves Town A traveling eastbound at 8 miles per hour toward Town B, which is 200 miles away. At the same departure time, Train B leaves Town B traveling westbound toward Town A at 60 miles per hour. When do the two trains meet?[17]

And here's an example of an insight question: A boy and his father are in a car accident. The father is killed. The boy is rushed to the hospital and taken into an operating room. The surgeon walks in and says, "I can't operate on this boy. He's my son!" How is this possible?[18]

Both chronotypes did better on analytical problems when they were on-peak and alert — morning types in the morning, evening types in the evening. Both chronotypes did better on insightful problems at off-peak times, when the mind was distractible and groggy — morning types in the evening, and evening types in the morning.

What this means for planning a spectacular wedding or fabulous vacation: Envision it and brainstorm when you're tired and off-peak.

And what about turning those visions into a reality? **The logistics rhythm** is called for when you're doing the grunt work of planning the hotels and flights, booking the caterer and florist. Do your research and price comparisons when you are at your analytical best at on-peak alert hours.

When you and your partner, friend, or family member discuss visions and logistics, **the attention rhythm** comes into play. Hashing over the endless details requires what psychologists call sustained attention, or keeping your thoughts focused on boring logistics when you'd rather do just about anything else.

When are you best able to pay attention and stay on task for the longest period possible? In a 2014 Spanish study,[19] subjects were sorted by chronotype and then tested on their ability to perform tasks at different times of day. Morning types did well in the morning session and in the evening, and showed stable and accurate performance during optimal times. Evening types did better in the evening session than in the morning, and weren't able to hold onto their concentration for as long as morning types at any time of day. As suspected, Wolves are less conscientious—less inclined to stay on task—than Lions.

The study also found that both types' accuracy and responsiveness declined over time. Even Lions get distracted if they've been hammering away at a task for too long. In a real-life situation (outside the lab or study context), it's wise to limit logistical discussions to forty-five minutes. If you haven't finished the planning in that time, put the work aside until the next day for another forty-five-minute session, and so on until the job is done. Also, keep in mind that, according to the Spanish study and my clinical observations, every task you set out to accomplish is easier after a decent night's sleep.

The Rhythm Recap

The vision rhythm: When your brain is best equipped to fantasize and come up with creative ideas.

The logistics rhythm: When your brain is best equipped to analyze the options, research, figure out budgets, and make reservations.

The attention rhythm: When your chronotype can sustain attention to detail.

THE WORST TIME TO PLAN SOMETHING IMPORTANT

At the last minute.

THE BEST TIME TO PLAN SOMETHING IMPORTANT

Since each chronotype excels at different tasks, it might be smart to divide jobs. In general terms, Dolphins and Lions should take charge of the research and planning. Bears and Wolves should take charge of the vision stuff. That said, here are the best times for each chronotype to do both:

Dolphin: Talk about ideas: **8:00 a.m. to noon.** Research and solidify plans: **8:00 p.m. to 10:00 p.m.**

Lion: Talk about ideas: **8:00 p.m. to 10:00 p.m.** Research and solidify plans: **6:00 a.m. to 9:00 a.m.**

Bear: Talk about ideas: **2:00 p.m. to 3:00 p.m.; 6:00 p.m. to 9:00 p.m.** Research and solidify plans: **10:00 a.m. to 2:00 p.m.**

Wolf: Talk about ideas: **8:00 a.m. to noon.** Research and solidify plans: **6:00 p.m. to 10:00 p.m.**

TALK TO YOUR KIDS

Failure: Talking to little brick walls; being ignored or rebuffed despite your efforts to get the kids to open up.

Success: Engaging your children in conversations when they're open to it, catching a glimpse of their state of mind, helping them, and cementing your bond.

The Simple Science

When should you talk to your kids? As a psychologist, my simplest answer is: "Whenever they choose to talk to you."

It's only too common to ask a child about his day and get "nothing" and "whatever" in return. But then, when you least expect it, they reveal fascinating insight into their inner and social lives. Nine times out of ten, those golden moments happen when you're busy doing something else. Children have uncanny timing for seeking out parental attention when you are least available to give it.

When you initiate conversations with them, check your bio-timing. When you sit down for little talks with kids, "when" is more important than what you have to say if you want them to listen.

The distraction rhythm. Children are most open to conversation when they're distractible (not focused on a task, be it homework, a computer game, or Snapchat) and at low energy (off-peak and tired, as opposed to bouncing off the walls). And when would that be? It depends on the child's age.

Adult populations are divided by the four chronotypes. But children are far more likely to fall into one or another category depending on their age. For example:

- **Infants** are majority Wolves, sleepy during the day and active overnight.
- **Toddlers through kindergartners** are majority Lions, waking pre-dawn and then falling asleep early. They benefit greatly from afternoon

naps that allow their bodies to recharge. If a toddler or preschooler doesn't sleep on a Lion's schedule, it's probably because he's habituated to his parents' schedule or takes long afternoon naps.

- **Grade schoolers** are majority Bears. They wake and fall asleep on a solar schedule and fade out of afternoon napping.
- **Teenagers** are majority Wolves. They are zombies in the morning, with energy surging at night, much to the annoyance of parents who want peace and quiet overnight.

To determine when a child is primed for conversation, consider age more than any other factor.

- **Ages one to six:** Initiate important conversations immediately after lunch and dinner. Little Lions will experience a drop in blood sugar known as the postprandial dip after eating. For about thirty minutes, they slow down but don't shut down, giving you the perfect window for conversation.
- **Ages seven to twelve:** Initiate talks after school from 3:00 p.m. to 5:00 p.m. during their predictable late-afternoon mood boost. I've often advised patients to talk to their kids in the car while driving them to classes and sporting events. The combination of the bio-timing, the side-by-side (not face-to-face) dynamic, and forced enclosure works wonders. But if they have a bad day at school, switch to their topic, and don't persist with yours.
- **Ages thirteen to eighteen:** Initiate conversations around 10:00 p.m. Teen Wolves are downright gabby late at night. If you can catch them within an hour of bedtime, they'll surprise you with how much they're willing to reveal.

Now you know when they'll be open to talking to you. But conversation is a two-way street. When are **you** in the best form to talk to them?

The patience rhythm is when you can keep your cool—for your sake and your child's. According to a 2014 University of Pittsburgh study[20] of 976 thirteen-year-olds, parental yelling was as devastating as physical abuse, resulting in a child's social misconduct and depression. Parental warmth after the fact didn't mitigate the effects of yelling, either. There is a direct link between your mood when you are talking to a child and that child's ability to talk to anyone else in the wider world.

Talk to your kids when cortisol levels are low (mid-afternoon or at bedtime), serotonin level is high (after carb-heavy meals or post-exercise), and oxytocin is flowing (after sex or some other kind of affection).

The Rhythm Recap

The distraction rhythm: When a child is off-peak, can't focus, and is a bit tired.

The patience rhythm: When you are best able to keep your cool.

THE WORST TIME TO TALK TO YOUR KIDS

7:00 a.m. and **5:00 p.m.** At 7:00 a.m., everyone is rushing around. The kids are hungry and hyper-alert with their morning cortisol level uptick (unless they are teen Wolves—in which case you are simply happy they managed to put on their clothes correctly). You experience the same blood pressure increase, as well as being easily irritable with sleep inertia fog. Obviously, you will talk logistics at breakfast. Save advice and probing for another time.

Another rush hour is 5:00 p.m. The kids are hungry and cranky, with homework still ahead. You're stressed out after the workday and still have many chores to do. In that state of mind, attempting an in-depth conversation will be frustrating and unproductive.

THE BEST TIME TO TALK TO YOUR KIDS

The underlying question is, "When is your patience peak?"

Dolphin: 7:00 p.m., after a carb-heavy dinner and before your nightly cortisol level uptick.

Lion: 3:00 p.m. If you can't talk to them at breakfast, grab them after school. You will hit a patience wall by 8:00 p.m. or 9:00 p.m.

Bear: 4:00 p.m. Your patience peak is 4:00 p.m. on Sunday, after you've had morning sex, a big lunch, and a long walk. Otherwise, check in after school.

Wolf: 8:00 p.m. Wolves are the bedtime parents, the ones who aren't too tired for homework help and bedtime reading. Take advantage of your elevating mood in the evenings.

"TURN THAT THING OFF ALREADY!"

I've advised adult Dolphins, Lions, Bears, and Wolves alike to shut off screens an hour before bedtime to ensure a higher quality of sleep. The Power-Down Hour is even more important for adolescents. Teenagers already have a tendency toward Wolfishness, and Internet use late at night exacerbates it. Eveningness in adolescents is associated with poor school performance,[21] anxiety, depression, Internet addiction, and perceived lack of family support.[22] As much as they might resent you for enforcing a house rule to turn off screens at a reasonable hour, adolescents will benefit from the practice academically and emotionally. They'll thank you later in life (well, probably not, but we can hope). Check out more on the Power-Down Hour at www.thepowerofwhen.com.

Fitness

GO FOR A RUN

Failure: Forcing yourself to run, hating every minute of it, being embarrassingly slow, risking injury, or chronically blowing off the activity.

Success: Running like the wind, sticking with it, losing weight, and feeling great while you're doing it.

The Simple Science

The optimal time to take a run depends on your specific goals. Are you running for weight loss? To compete in a race, or against yourself? To improve your overall health?

To follow **the fat-burning rhythm,** you can run in the morning or in the evening. A fasting workout within a half hour of waking converts fat into energy because you haven't ingested any carbohydrates yet. If you are inclined to run before breakfast, make sure you're well hydrated. Afterward, eat a breakfast of 50 percent carbs and 50 percent protein to keep the metabolic fires stoked. Evening workouts give the body an endorphin boost that decreases appetite at the end of the day, when people are susceptible to pigging out. But studies show that morning exercise is more likely to become a habit, since you don't have all day to talk yourself out of it.

For **the performance rhythm,** run according to your preferred wake-up time. Numerous studies have confirmed that people run faster, cycle faster, and hit the baseball harder in the afternoon and evening than in the morning. However, a 2015 British study[1] proved that the most significant factor in predicting athletic peak performance across a wide range of sports is the time athletes prefer to rise relative to the time they perform. Researchers had athletes train at several times through a day and measured their speed and agility. Early risers performed best in the late morning. Intermediate risers did best in the afternoon. Late risers did best in the evening. An individual athlete's performance varied by as much as 26 percent from morning workout to evening workout. If you are a Lion and your event is scheduled for the early morning, you are in luck. Otherwise, you'll be playing at a distinct disadvantage. Or, you can adjust your bio-time as if you've flown to a new time zone to get an optimal running performance. (See "Travel," page 313.)

Deep sleep in and of itself improves your immune function and cardiac health, lowers your blood pressure, and relieves anxiety. Anything that deepens sleep is a bonus for overall health. To improve the quality of your sleep for a healthy **rest rhythm,** run in the morning. In an Appalachian State University study,[2] researchers had groups of subjects walk on a treadmill at three different times—7:00 a.m., 1:00 p.m., and 7:00 p.m.—and monitored their blood pressure and sleep. The 7:00 a.m. group had a 10 percent drop in blood pressure post-workout and a 25 percent drop later that night, along with a 75 percent increase in deep delta wave sleep—significant improvements compared to those experienced by the 7:00 p.m. walkers.

The Rhythm Recap

The fat-burning rhythm: When to run to burn fat and boost metabolism.

The performance rhythm: When to run for faster times.

The rest rhythm: When to run for higher quality of sleep.

THE WORST TIME TO GO FOR A RUN

6:00 a.m. Running at dawn increases injury risk. Core body temperature is low, and muscles and joints are susceptible to strain and tear. If you can wait until ninety minutes after waking, your temperature will be higher, and injury risk drops significantly. This changes with the seasons, especially if you are living in a dry area like Arizona. (For more information on how the change of seasons affects chronorhythms, go to "Chrono-Seasonality," page 321.)

THE BEST TIME TO GO FOR A RUN

Lion: 5:30 p.m., for faster times and an energy boost.

Dolphin: 7:30 a.m. Morning runs help you sleep longer and more deeply, and you especially benefit from the restorative delta waves that light sleepers need more of.

Bear: 7:30 a.m. or **12:00 p.m.,** for either a fasting pre-breakfast fat-burn or the afternoon appetite suppressant.

Wolf: 6:00 p.m. for performance and an evening fat-burn.

"I'M NOW A PART OF ALL THIS ACTIVITY AND LIFE"

"At the end of the workday, exercise was the last thing I wanted to do," said **Robert, the Lion,** who lives in Boston. "But I forced myself to try it. I brought my workout clothes to the office and started running home instead of driving or using mass transit. The first thing I noticed was that the sidewalks were busy with a lot people, jogging, walking dogs, and pushing strollers. It was completely different from being alone on the road at dawn, seeing one or two people. Sometimes the crowds can be annoying. But I've found that I like the company and the feeling that I'm now a part of all this activity and life. I get home, take a cool shower, and find that I do have more energy to meet friends for dinner. It definitely took some adjustment. But it was worth it. A couple of my colleagues got the 'run home' bug, so we do it together. I don't know if it's the friendly competition or having a warmed-up body, but I'm faster, too."

PLAY A TEAM SPORT

Failure: Embarrassing yourself on the field, being a bad sport, and/or counting the minutes until the game is over.

Success: Playing well, having fun, and behaving in a way that would make your mother proud.

The Simple Science

People who play team sports have been found to possess "mental toughness,"[3] a collection of qualities that includes resilience (the ability to bounce back), perseverance (the ability to carry on), and optimism (the ability to see a positive outcome). Being on a sports team helps kids develop essential skills like self-efficacy (belief in yourself) and emotional intelligence (the ability to read and react appropriately to others' feelings). Plus, getting together with friends, doing something physical outside, and sharing the thrill of victory (and the agony of defeat) is just a fun way to spend a weekend afternoon. Any kind of group pursuit (such as book clubs, card games, video games, etc.) will yield similar mental and emotional benefits. The beauty of team sport is that you get physical benefits as well.

To have the most fun, get along with your teammates and opponents, and knock it out of the park, remember one time-wise tip: Play at dusk.

Just how into the game are you? Do you stand in the outfield and count the minutes until the inning is over so you can sit in the dugout and have another beer? Or do you really care about winning and advancing to the playoffs? Certain chronotypes tend to be more competitive and goal-oriented than others. In a soccer match between the Lions and the Wolves, the Lions would handily win, since the Wolves probably wouldn't show up (unless the game is called for 9:00 p.m.).

Regarding **the competitiveness rhythm:**

- **Dolphins'** productive, energetic time is from 4:00 p.m. to 6:00 p.m., but playing team sports isn't a common pastime for them. Insomniacs are just too tired to play.

- **Lions** are more likely to participate in team sports[4] and to be aggressive and ambitious on the field. If they play in a night league, they'll need late-afternoon exposure to sunlight and a high-protein snack to keep them motivated.

- **Bears** reach their "bring it" peak midday and will play hard until they hit a stamina wall at nap time. They'll rebound in the early evening.

- **Wolves**...the only sport I can really see them playing in a group is beer pong. Evening types are just not into team sport participation and are more sedentary in general.[5] Watching sports, on the other hand, can be Olympic with them. The majority of athletic Wolves in my practice choose a solo sport like running or swimming. If a Wolf did join a league, she'd prefer to play night games.

The coordination rhythm, or when you'd be least likely to make a fool of yourself in front of all your friends, correlates with how many hours have passed since you woke up, how tired you are, and how difficult the game is to play.[6] If you're tired and it's been hours since wake time, you might be clumsy. But if you're alert, fresh, and know what you're doing, your coordination will impress spectators and teammates. For most people (Bears), hand-eye coordination peaks between 5:00 p.m. and 8:00 p.m. Lions: 3:00 p.m. to 6:00 p.m. Wolves: 6:00 p.m. to 9:00 p.m.

The power rhythm, or when you are at your strongest and fastest, can be tracked along with fluctuations in body temperature. When your temperature is higher, you'll have greater lung capacity, blood flow to muscles, and flexibility. Your reflexes are faster and you'll have raw power in your arms, legs, and back. Strength and stamina peak in the evening between 6:00 p.m. and 9:00 p.m. for Bears (shift this range two hours earlier for Lions and an hour or two later for Wolves). If you're into racquet or stick sports — tennis, racquetball, golf, baseball, hockey — you'll have stronger grip strength in the late afternoon / early evening, too.

Ever wonder why professional sports games are scheduled for the evening? It's not only for prime-time TV ratings. It's because the athletes perform at a higher level at that time. Well, some of them do, depending on chronotype. In 2011, researchers at the Martha Jefferson Hospital Sleep Medicine Center, in Charlottesville, Virginia, analyzed two years' worth of the batting averages of sixteen Major League Baseball players from seven teams. They also asked each player to fill out the Morningness-Eveningness Questionnaire (MEQ). Nine players were evening types, and seven were morning types. For the morning types, the highest batting average (.267) was when game time was at 2:00 p.m., and the lowest (.252) was at 8:00 p.m. game times. The opposite was true for evening types. Their averages were highest (.306) when game time was at 8:00 p.m. and lowest (.259) at 2:00 p.m. games.

The sportsmanship rhythm is all about mood. When you are in a good mood, you're less likely to go nuclear over an ump's bad call or intentionally foul an opponent. Recreational team sport participants at their positive emotional peak are just more fun to play with. Most of us reach our emotional mellow point in the later afternoon/early evening.

The Rhythm Recap

The competitiveness rhythm: When you are most motivated to win.

The coordination rhythm: When your hand-eye reflexes are sharpest.

The power rhythm: When you are stronger, faster, and most flexible.

The sportsmanship rhythm: When you'll be a good sport on the field.

THE WORST TIME TO PLAY TEAM SPORTS

Early morning. Even Lions aren't warmed up enough in the early morning for competitive or team sports and run a higher risk of injury. Cortisol levels and testosterone blood concentration are high then, and participants might be too amped up to relax and have fun.

THE BEST TIME TO PLAY TEAM SPORTS

Dolphin: 5:00 p.m. to 7:00 p.m. The time of day when you are least tired.

Lion: 2:00 p.m. to 4:00 p.m., the nexus of your strength, mood, and coordination.

Bear: 6:00 p.m. to 8:00 p.m., the nexus of your strength, mood, and coordination

Wolf: 6:00 p.m. to 9:00 p.m., the nexus of your strength, mood, and coordination.

PRACTICE YOGA

Failure: Going through the motions, possibly pulling something.

Success: Moving deeply through the poses to improve fitness, mood, and cognition.

The Simple Science

Yoga has been around for thousands of years for a reason: It works. If you regularly practice it, you'll become stronger and more flexible physically and mentally. Deep breathing expands your lung capacity. The introspection of yoga can give you a healthy perspective on your problems and stressors. Yoga poses can help quiet the parasympathetic nervous system, making you relax. Stretching and bending will expand your range of motion and muscle tone.

When is the best time of day to do downward-facing dog, cobra, and cat-cow?

If your goal in doing yoga is to improve your range of motion and muscle tone, follow **the flexibility rhythm**, or when your body is supple and able to move deeply into each pose. You are most flexible when your body temperature is at or near peak. Think about it: A warmed-up body is loose and ready to perform. The warmth comes from an increase in temperature. Fans of Bikram or Prana Power yoga know that they're more flexible in a hot environment. This is why you see more and more hot yoga studios. The same warm-up principle applies to the internal environment of your own body. Your temperature will be sufficiently warm for yoga three hours after waking and again in the early evening. So sign up for pre-lunch or after-work class.

A cool body is stiff. When your body temperature is low, you'll be stiff, and this is the time you're most likely to sustain an injury doing any type of physical activity. Temperature is low within ninety minutes of waking, in the mid-afternoon, and in the evening three hours before bedtime. So avoid sunrise and post-lunch classes and the last session of the night.

Or, if you're not into backbends, and practice yoga just to relax, do your poses on **the relaxation rhythm.** There is a wealth of scientific proof that deep breathing and yoga stretching lowers cortisol level and blood pressure.[7] It's time-wise to practice yoga to counteract stress as needed throughout the day. I encourage anyone to incorporate low-intensity stretching into their nightly Power-Down Hour before bed, and to use deep breathing techniques whenever anxiety strikes. Go to www.the powerofwhen.com for video instructions.

Along with the physical benefits, yoga improves mental abilities, too, such as mindfulness, alertness, concentration, and memory. A 2014 study done at the University of Chieti-Pescara in Italy[8] looked at yoga's **mind-body connection rhythm.** We know that certain personality traits and learning behaviors correspond to different chronotypes. The research-ers observed that yogis tend to have certain personality traits and learning preferences. They tested 184 yoga instructors on personality, morningness-eveningness preference, and thinking style. There were relatively few Wolf yogis in the study (8 percent among the study sub-jects, compared to 15 percent in the general population), and there was a preponderance of Bears (71 percent of the subjects, vs. 50 percent of the general population). Lions made up 20 percent of the subjects—which matches the proportion of Lions in the general population. Naturally, the Lion yogis scored higher than Bears and Wolves in conscientiousness, scrupulousness, and emotional stability. The Wolves scored higher in impulsivity and were more neurotic and more sociable. No surprise there. What did make me look twice was that the vast majority of yogi trainers of each chronotype—including the Lions—scored high in openness and right-brained creative thinking. Does doing a lot of yoga make any chronotype more open and insightful? The research suggests that, yes, it does!

Yoga also makes people (even Wolves) more optimistic. A 2014 study from Chicago State University[9] set out to track the mood changes in the different chronotypes after participants had done a single hatha yoga ses-sion. All groups reported a brighter outlook post-yoga compared to the control group, whose members had listened to a lecture. However, the eveningness group self-reported the most significant positive shifts. Since Wolves tend to be pessimistic, the results suggest that a yoga practice could be life-changing for those in need of a healthy way to lift their spirits.

The Rhythm Recap

The flexibility rhythm: When your body is the most flexible.

The relaxation rhythm: When you can use exercise, especially yoga, to reduce cortisol level and lower blood pressure.

The mind-body rhythm: When exercise, especially yoga, can change your outlook and increase mindfulness and optimism.

THE WORST TIME TO PRACTICE YOGA

Sunrise. I know sunrise yoga is popular, but even if you're a Lion, I don't recommend it. Your body is not warmed up enough for deep stretching, and you could injure yourself. Only do this after you consider yourself at intermediate level.

THE BEST TIME TO PRACTICE YOGA

Dolphin: 10:00 p.m., to lower elevated cortisol level and blood pressure at night.

Lion: 8:00 a.m. or **5:00 p.m.** Right before work, or right after.

Bear: 12:00 p.m. or **6:00 p.m.** Pre-lunch or sunset yoga.

Wolf: 6:00 p.m. or **10:00 p.m.** Pre-dinner or for relaxation during the Power-Down Hour.

TRAIN FOR STRENGTH

Failure: Sporadic strength training that doesn't increase muscle mass, improve muscle tone, or speed up metabolism.

Success: Regular strength training that increases muscle mass, improves muscle tone, and speeds up metabolism.

The Simple Science

Building muscle mass not only increases your strength, but it also speeds up your metabolism. The more lean muscle mass you carry, the easier it is to burn fat for energy. Any fitness regimen to burn fat, speed up your metabolism, and improve muscle tone should include some form of resistance training—lifting weights or isometric exercise—that forces large muscle groups to contract against an external force, like weights or the ground (push-ups and planks are resistance training methods).

What's the **the muscle growth rhythm?** In a 2009 study[10] from the University of Jyväskylä, Finland, researchers set out to determine the impact of time of day on hypertrophy (muscle mass gains) in men. A group signed on for twenty weeks of daily randomized training sessions, either in the morning (between 7:00 a.m. and 9:00 a.m.) or the afternoon (between 5:00 p.m. and 7:00 p.m.). Their muscle mass was measured with MRI machines every ten weeks at random times. The morning and the afternoon training groups *both* showed muscle volume increases of about 3 percent, leading researchers to conclude that time of day doesn't matter that much for growth. Whether you work out early or late, you're going to see similar results in hypertrophy over a three-month period—if you train *daily*. The control group that didn't exercise regularly saw no increase in volume.

Trainers who set goals for improvement should follow **the muscle strength rhythm**—that is, they should know when their clients are strongest. According to numerous studies, including research from the

University of West Scotland,[11] optimal resistance training strength occurs in the late afternoon, when core body temperature is high and testosterone and cortisol concentrations are low. You'd think that muscle strength would be greatest when testosterone is elevated (in the morning). Actually, it's not the concentration of testosterone that matters but the ratio of cortisol to testosterone, or the C/T ratio.[12] The ratio of C to T that is most favorable to strength performance is **in the afternoon for Lions, the early evening for Bears, and the late evening for Dolphins and Wolves.** Since your body temperature is highest in the evening, too—with increases in heart rate, muscular blood flow, joint flexibility, and glucose metabolism—your body is primed to perform and you'll be able to lift more pounds, or hold a plank for longer.

I don't personally do much strength training, but from what I've observed watching weight lifters at the gym, specifically their strained facial expressions, it appears to be very painful. This brings up the question of circadian rhythm and pain. What is your **pain tolerance rhythm**? A University of Warsaw, Poland, study[13] tested sixteen healthy morning types and fifteen healthy evening types by applying heat stimuli to their wrists nine times throughout the day. Both groups' pain tolerance changed significantly from morning to night, but the morning types could take more heat than evening types all day long—50 degrees Celsius vs. 47 degrees Celsius (or 122 degrees Fahrenheit vs. 116 degrees Fahrenheit). **This data corroborates the abundance of research about Wolves' aversion to exercise.** They just can't stand the pain.

The Rhythm Recap

The muscle growth rhythm: When to exercise to increase muscle volume.

The muscle strength rhythm: When your cortisol to testosterone ratio is most favorable for lifting weights.

The pain tolerance rhythm: When different chronotypes can withstand pain.

THE WORST TIME TO TRAIN FOR STRENGTH

6:00 a.m. Body temperature is lowest in the early morning. If you train when you're cold, you are susceptible to injury. Muscular blood flow, joint flexibility, and hormonal ratios are unfriendly to resistance exercises and weight lifting.

THE BEST TIME TO TRAIN FOR STRENGTH

Dolphin: 8:00 p.m. Your temperature rises and your cortisol level and heart rate increase at night, ideal for muscle growth and strength.

Lion: 2:30 p.m. to 5:00 p.m., your muscle strength window.

Bear: 4:00 p.m. to 7:00 p.m. Your body is primed, but if you don't exercise every day, you won't increase muscle volume, which, in turn, speeds up metabolism.

Wolf: 6:00 p.m. to 7:00 p.m. Routine matters for increasing muscle volume and speeding up metabolism. Although your strength peak comes later, it's unlikely you'll actually exercise at 10:00 p.m. Also, exercise within three hours of bed can delay sleep.

CHAPTER 8

Health

FIGHT ILLNESS

Failure: Compromising your immune system, making it easier to become ill and harder to get well.

Success: Boosting your immune system to prevent illnesses and fight them off faster.

The Simple Science

The science of immunity is not simple, actually. It's complex, involving proteins, blood cells, antibodies, hormones, and receptor cells. The basic facts that everyone needs to know:

• Your immune system is made up of armies of white blood cells and antibodies that patrol your body in search of bacteria, infection, inflammation, and malignancies.

• Your immune system — like every other system in your body — is affected by circadian rhythms. Treating illnesses on bio-time is more efficient and effective.

In the last few years, scientists have discovered the bio-time of your body's defenses. Immune function is most active at night, when we are resting, and less active during the day, when we are out and about. The bulk of healing goes on during the first third of a night's sleep, when your body is recovering from the busy day, and when you undergo physical restoration due to an increase in slow wave deep sleep.

However, your white blood cells — the front line of defense — keep their own bio-time, and it doesn't necessarily stick to the schedule. If white blood cells detect bacteria, infection, or inflammation somewhere inside you, they can go rogue (go on or off their usual bio-time) and attack the bad cells at any time throughout the day. **The immunity override rhythm** was identified by scientists at Trinity College in Dublin and the University of Pennsylvania.[1] They proved that **we have an alarm** inside each immunity cell that says, "Day or night, it's time to fight." Researchers are especially excited about using this insight to develop and improve the new class of "immunotherapy" drugs—custom-made antibodies and proteins that act as the "on" switch to activate our body's natural defenses to attack the invader, be it bacteria or a tumor, more aggressively.

Although it's been proven that our immune system kicks in when the sun goes down, it's also true that cancerous tumors grow more aggressively at night. **The growth rhythm** was identified by researchers at Israel's Weizmann Institute of Science in 2014. Researchers examined[2] the link between cell growth and a class of steroid hormones called glucocorticoids (GCs). GCs control metabolism, energy, and alertness. They peak during the day, decrease at night, and interact with epidermal growth factor receptor (EGFR) molecules. An EGFR is exactly what it sounds like—a receptor on the surface of a cell that, when unlocked, allows the cell to grow and divide, which is all well and good in healthy, normal cells. EGFRs also cause malignant (cancerous) cells to grow and divide.

In mice, when GC levels went down, EGFR involvement went up in tumors, making them increase and spread faster. Researchers next administered EGFR-inhibiting drugs at different times of the day to see if bio-time could slow tumor growth in combination with drug treatment. **Timing *did* make a huge difference. Tumors in mice treated during sleeping hours were smaller than those treated during the day.** In the future, it's likely that EGFR-related cancers will be treated when GC hormones are low—to enhance their effectiveness.

The sleep duration rhythm, or getting enough hours of sleep per night, has been proven to decrease susceptibility to minor illnesses such as the common cold.[3] In a 2015 study by researchers at the University of California at San Francisco and Carnegie Mellon University, 164 adults were given wrist monitors to track their sleep duration for a week. Then the subjects were moved to a lab, given nasal drips with the cold virus, and tracked for another five days. The subjects who slept less than six hours were significantly more likely to get sick than those sleeping seven hours per night. All other variables—gender, health practices, BMI, psychological issues—were ruled out. One hour per night made all the difference.

The sleep disruption rhythm—fragmented sleep or sleeping out of sync with circadian rhythms—might be even more damaging in terms of disease susceptibility than low sleep duration. In a 2014 study,[4] researchers at the University of Chicago corralled mice into two groups. The first group's subjects endured a week of fragmented sleep in the worst way: A motorized brush ran through their cage, waking them up periodically. The other group was undisturbed and slept peacefully. Both groups were injected with tumor cells. Four weeks later, the mice were examined. The fragmented sleep group's tumors were double the size of, and far more invasive than, the well-rested mice's tumors. The mushrooming effect was caused not by the cancer cells themselves but by the compromised immunity of the sleep-disrupted mice. Instead of attacking malignant cells in the tumor core (as in the case of the well-rested mice), the immune system of the sleepy mice misdirected white blood cells to blood vessels around the tumors' edges, promoting rapid growth. The good news: If a certain protein that signals tumor growth—TLR4, or toll-like receptor 4—was blocked, tumor growth was kept in check, even in sleep-deprived mice. So there is hope for new avenues of treatment by blocking this protein, as well as getting the word out that quality sleep does a lot more than make you feel refreshed. It could save your life.

The Rhythm Recap

The immunity override rhythm: When cells switch on their disease-fighting capabilities, day or night.

The growth rhythm: When cells are best able to fight the expansion of tumors.

The sleep duration rhythm: When you don't get enough sleep to fight illness.

The sleep disruption rhythm: When you don't get high enough quality sleep to fight illness.

THE WORST TIME TO FIGHT ILLNESS

2:00 a.m. Immunity improves at night, but only if you're actually asleep! If you are awake at this hour, you're not getting enough high-quality sleep, which could compromise your immune system.

THE BEST TIME TO FIGHT ILLNESS

Getting sufficient quality sleep will do as much to fight and prevent disease as quitting high-risk behaviors such as smoking, drinking, and eating junk food. Shoot for seven hours per night. An extra hour can make all the difference.

Dolphin: 11:30 p.m. to 6:30 a.m.

Lion: 10:30 p.m. to 5:30 a.m.

Bear: 11:30 p.m. to 7:00 a.m.

Wolf: 12:30 a.m. to 7:30 a.m.

HELP DOCTORS HELP YOU

Some time-wise tricks for interacting with doctors and pharmacists:

- **Schedule checkups in the morning.** Your wait will be shorter, sparing you annoyance, irritation, and hunger if you're giving a fasting blood sample. Stiffness and lung function are poor in the morning, too, so arthritis and asthma are at their worst. Your doctor can better assess your care needs then.
- **Schedule surgery in the morning.** The effects of anesthesia are on bio-time. You're more likely to suffer side effects from being put under in the afternoon, according to a Duke University Medical Center study of 90,000 surgeries performed at Duke University Hospital from 2000 to 2004. The study's patients also reported more administrative problems in the afternoon—waiting around, delays in paperwork, the kind of stuff that sends blood pressure skyrocketing. What's more, surgeons made fewer errors in the morning. Adverse incidents happened least often

during on-peak alertness hours, between 9:00 a.m. and 12:00 p.m. Mistakes occurred most often during off-peak alertness hours, between 3:00 p.m. and 4:00 p.m. Patients were more likely to feel post-op pain and nausea in the afternoons as well.

- **Call in prescriptions in the morning and pick them up in the afternoon.** The large window gives the pharmacist plenty of time to fill prescriptions and prevents mistakes caused by rushing.

GET A FLU SHOT

Failure: Get the shot when it's least effective and most painful.

Success: Get the shot when it's most effective and pain-free.

The Simple Science

Every year, it seems that the flu shot takes some hits from the media. News sources report that the vaccine blocks the virus only half the time or that it only reduces the severity of the illness. A vaccine's potency is about its ability to create antibodies—the specific blood protein that attacks a specific antigen, like the H1N1 virus. For 50–70 percent of the people who get it, the flu vaccine—a tiny dose of the virus itself— triggers the body to produce enough antibodies to fight off the full-blown illness. The rest of those who get the vaccine don't produce enough anti- bodies to make a difference. Hearing the statistics, many choose not to bother getting the vaccine: Why make an appointment, go over there, and pay the $20 if it's not going to work anyway?

If there is even a 10 percent chance of not getting the flu and spending two weeks in bed in agony, the shot is worth the time and money. I rec- ommend that everyone over six months of age get an annual flu shot. It's a necessary evil for your overall health care routine. And, if you're going to do it, you might as well do it on bio-time to maximize efficiency and minimize pain.

This book is about making minor adjustments to your daily sched- ules, but in this case, time of year is important, too. **The seasonal rhythm** for the flu virus, according to the Centers for Disease Control and Pre- vention, the government agency that has charted the influenza virus sea- son from 1982 through 2014, is from October to March, peaking in December to February. Since the flu shot takes two weeks to become effective, getting the shot in January would be like closing the barn door after the horse has already run out. **The best time to get the shot is in early October,** at the very start of the season, as soon as the vaccine becomes widely available. When I notice a lot of Party City ads for Halloween costumes on TV, I think, *Time for a flu shot.*

The pain sensitivity rhythm isn't a major concern for most of us. The shot itself causes minimal pain, plus a bit of soreness on the arm for a day or two. No big deal. (It is true, however, that some kids will scream in anticipatory agony just seeing the needle.) And studies have found that pain is on bio-time. So in fact it is possible to reduce any discomfort that might be associated with shots. In 2014, scientists at the University of Haifa[5] in Israel inflicted intensities and types of pain—mechanical, heat and cold—on their forty-eight male subjects to test their tolerance in the morning, afternoon, and evening. **The subjects could withstand pain for longer periods and with greater intensity in the morning.**

Is there **an immunity rhythm** that would suggest when to get the shot so that it has the best chance of protecting you from illness? As mentioned above, if the vaccine causes you to produce enough antibodies, it'll protect you from illness. So when does the body naturally produce antibodies at a faster rate? Fifteen minutes before exercise.

In 2011, researchers at Iowa State University including Marian Kohut, a professor of kinesiology, gave a group of students the H1N1 flu vaccine and then instructed some of them to run or ride a stationary bike for ninety minutes immediately after. The control group didn't exercise. The subjects were tested a month later to count how many antibodies they produced. The exercise group had double the number of antibodies as the sedentary group. She followed up that study on mice and found that the length of time spent exercising made a difference as well. Too long (three hours) or too short (forty-five minutes) didn't produce results that were as impressive as those experienced by subjects after they had exercised for ninety minutes. Why does cardio help produce more antibodies? Exercise increases circulation. So getting the blood moving right after receiving the shot helps to spread the vaccine all over your body, increasing widespread production of antibodies.

An English study[6] tested the impact of subjects weight lifting *before* receiving a vaccine. The University of Birmingham researchers had twenty-nine men and thirty-one women do biceps curls on their nondominant arm to 85 percent of their maximum capacity; then, six hours later, the participants received the flu shot in the same arm. Eight weeks later, the subjects were tested. Interestingly, the female curlers did have an increase in antibodies compared to the control group, but not as much of an increase as the male participants. The cause could be inflammation in the arm muscle that prevented the vaccine from getting to the rest of the body quickly. In any case, for both genders, if you pump iron six hours before your shot and then ride a stationary bike or take a brisk walk for ninety minutes afterward, you'll be in great shape for flu season. So if your gym has a flu shot station, you may want to check it out!

The Rhythm Recap

The seasonal rhythm: When, on a month-to-month basis, to get a flu shot.

The pain sensitivity rhythm: When you are most sensitive to pain.

The immunity rhythm: When your body will produce antibodies at a faster rate.

THE WORST TIME TO GET A FLU SHOT

A lazy afternoon or evening in January. It's simply too late.

THE BEST TIME TO GET A FLU SHOT

All chronotypes should get a flu shot in early October.

Dolphin: 1:00 p.m. Get the shot, and then take a long walk for immunity and an energy boost when you need it.

Lion: 4:45 p.m., right before a longer than usual workout.

Bear: 11:30 a.m., and then take a long walk to buy lunch and carry it back to work.

Wolf: 5:45 p.m. Wear exercise clothes to the pharmacy or doctor's office, and then walk, jog, or bike home, taking the long way.

GET A MAMMOGRAM

Failure: Not getting an accurate screening despite the pain and hassle of getting it.

Success: Getting an accurate screening without too much pain and annoyance.

The Simple Science

Mammograms save lives. We can all name someone we know who got an annual mammogram and is alive today because of it. Early detection of breast cancer goes a long way to helping you survive the disease. Five or ten minutes of discomfort are worth it to safeguard your health. Ideally, you'll get a clear, accurate screening without too much hassle and walk out of there happy and relieved to have it over with for the year. A few important "whens" to consider:

The comfort rhythm. Many women are unaware that caffeine increases breast tenderness and exacerbates the pain of fibroids. In a Duke University study[7] of 113 women with fibrocystic breast disease, two-thirds who substantially restricted caffeine for a year reported reduced or eliminated breast pain. Before a mammogram, cut back on caffeinated coffee, tea, and soda for a day or two to make the process easier to take. If you are addicted to coffee, schedule a mammogram before your first cup. Hormonal changes over the course of the month affect your pain response as well. Women are most sensitive the week *before* their period, when production of naturally pain-killing hormones like serotonin and endorphins drops.

What about **the convenience rhythm,** or when to schedule your mammogram appointment to get in and out quickly? If at all possible, claim the first or second screening of the day. As is true of all doctors' offices and screening facilities, schedules get more backed up and delayed as the day wears on. The earlier you get in, the less waiting around you'll have to endure.

Although pain and logistics are important, the main reason to have a mammogram is to spot a tumor. **The accuracy rhythm** is all about time of the month. According to a major study[8] by scientists at the Group Health Research Institute in Seattle, your best option is to schedule your appointment during the first week of your cycle (day one being the first day of your period). Analysis of 380,000 mammograms of women aged thirty-five to fifty-four over an eleven-year time span showed that there were positive readings (when a lump is found) 80 percent of the time during the first week. Breast tissue is less dense then—less swollen from water retention symptomatic of premenstrual syndrome, or PMS—and lumps are easier to spot. Later weeks of the cycle yielded positive readings approximately 70 percent of the time.

The Rhythm Recap

The comfort rhythm: When a mammogram will be least painful.

The convenience rhythm: When to schedule a mammogram so you don't have to wait around too long to get it.

The accuracy rhythm: When the screening will yield clear, accurate results.

THE WORST TIME TO GET A MAMMOGRAM

Late in the day, after drinking a pot of coffee, during the last week of your menstrual cycle.

THE BEST TIME TO GET A MAMMOGRAM

Shoot for the first appointment of the day during the first week of your menstrual cycle.

GO TO THE BATHROOM

Failure: Not going enough, going too often, straining to go, going at unpredictable times.

Success: Going at regular times with ease.

The Simple Science

The gastrointestinal (GI) tract is referred to, as we have seen, as your second brain. We've all experienced "gut feelings" or "gut instincts"—sensations that are, indeed, generated in our guts. Like our brains, our GI tracts are a nervous system—specifically, the enteric nervous system (ENS). The ENS is comprised of 100 million neurons in the lining of the alimentary canal—the long tube that starts in your esophagus and ends at your anus. All those nerve cells allow you to "feel" the digestive process and connect it to emotions and thoughts. For example, when you're anxious and have butterflies in your stomach, you're actually registering accelerated digestion caused by an increase of stress hormones and gut enzymes. The ENS controls the digestion and elimination processes without any help at all from your brain. Your GI tract has its own bio-time. Scientists call it the "gut clock."

Like the brain, your second brain produces neurotransmitters and hormones, the chemicals that signal to your cells to perform bodily functions. Certain hormones we associate with the brain are mass-produced in the GI tract, including 95 percent of your serotonin and 80 percent of your melatonin. Not surprisingly, drugs that inhibit or exaggerate first-brain hormones affect second-brain health. For example, serotonin reuptake inhibitors (SSRIs), which are used to treat depression, can cause irritable bowel syndrome. On the other hand, melatonin supplements wrongly used to treat insomnia can improve irritable bowel syndrome symptoms. It's all very complicated and interrelated. Suffice it to say, your gut is just as smart as your brain, if not smarter, and it knows exactly what it needs to do (or doo doo), and when.

Running a "gut check" on your gut rhythms is an excellent way to gauge your overall health.

As a hormone factory, the gut has quite the inventory, and many of the biochemicals it produces run on bio-time.[9]

- **Motilin** and **ghrelin** kick-start the digestive process of cascading muscle contractions, beginning in the stomach.
- **Gastrin, ghrelin, cholecystokinin,** and **serotonin** keep motility (the movement of food and waste) going through the small intestine and the colon.
- **Melatonin.** You may have thought it was only for sleep, but it has tremendous implications for the gut. It relates to hunger and satiety, and it controls the timing of the digestive process.

The impact of melatonin and the migrating motor complex — the 90- to 120-minute digestive cycle — sets off **the hormonal rhythm.** When it gets dark out and melatonin starts being secreted from the pineal gland, the GI tract and bowel function are suppressed. When the sun comes up and melatonin turns off — around 8:00 a.m. for Bears — the colon wakes up, too. Most Bears will have their first bowel movement of the day within ninety minutes of waking.

Regularity is a clear indication of colon health and, considering their interconnectedness, endocrine system balance as well. If you go at predictable times, your guts and your hormones are working harmoniously together. **The regularity rhythm** isn't only due to hormones, though. Consistent mealtimes — absolutely necessary to keep bio-time synchronized — plus a high-fiber diet and drinking at least six glasses of water daily will keep bowel movements predictable. You can plan accordingly, and you won't schedule an important job interview, for example, at the wrong time.

Nearly 30 percent of coffee drinkers have noticed the reliable and fast-acting **stimulant rhythm.** A cup of coffee can spur gut motility in just four minutes and makes you go within thirty, according to a British study[10] of ninety-nine healthy adult java junkies. Caffeine isn't responsible for the laxative effect, though. If it were, chocolate and soda would also bring on the urge. What does the trick is coffee's unique acidic quality and compounds that cause the stomach to speed up the digestive process and increase the body's production of two GI tract hormones, gastrin and cholecystokinin. Gastrin stimulates the muscle contractions of the intestines that move waste through the alimentary canal. Cholecystokinin signals the gall bladder and pancreas to release enzymes and bile to break down food fat and protein in the gut.

For some, bowel movement bio-time is synchronized to eating. **The reflex rhythm,** or gastrocolic reflex, is your body's response to ingesting food. After a meal, your stomach stretches, which sets off a hormonal chain reaction that triggers peristalsis, the muscle contractions that move stool through your intestines. Food comes in; waste, therefore, must go out. Most parents have noticed the gastrocolic reflex in action when a baby fills his diapers while breast feeding or drinking from a bottle. Adults need only excuse themselves from the table for a few minutes.

The Rhythm Recap

The hormonal rhythm: When the gut clock releases certain hormones, putting the digestive process into motion.

The regularity rhythm: When a combination of factors—fiber, water, and hormones—makes bowel movements occur at predictable times.

The stimulant rhythm: When the need to go coincides with drinking coffee.

The reflex rhythm: When the need to go coincides with eating.

THE WORST TIME TO GO TO THE BATHROOM

1:00 a.m. to 5:00 a.m. Melatonin is flowing in the middle of the night and should suppress your bowel activity. If you have to get up and go in the wee hours, there might be something wrong with your digestive or endocrine system.

THE BEST TIME TO GO TO THE BATHROOM

Dolphin: Whenever the urge happens. Constipation is an unfortunate side effect of insomnia.[11] My advice to insomnia patients is to eat fruits and vegetables at every meal, eat regularly three times a day, and not delay going for convenience or privacy—because that can compound constipation.

Lion: 7:00 a.m., an hour after waking.

Bear: 9:00 a.m., ninety minutes after waking. Remember to eat a high-fiber diet, eat at regular times, and drink a lot of water.

Wolf: 11:00 a.m. Wolves are slower to get started, colon-wise, and might not move for two or three hours after waking. Wolves who commit to eating breakfast will start to go earlier.

SEE A THERAPIST

Failure: Not seeing a therapist if you need help, or going when you're least likely to open up.

Success: Seeing a therapist if you need help, and going when you can open up.

The Simple Science

I'm a psychologist specializing in sleep medicine. The bulk of my work is helping my patients change their sleep habits to get more rest, but another big part of it is talking to them about their emotional issues. Talk therapy works. Whether you are going through a temporary crisis or a long-term problem, when you see a therapist could improve the quality of the sessions.

Certain chronotypes are more likely to need therapeutic help than others. **The life satisfaction rhythm**—whether you feel happy with your life and relationships and have a positive outlook for the future—leans decidedly in one chronotype's direction. Multiple international studies[12] have proven the link between morningness and high life satisfaction and, for that matter, eveningness and low life satisfaction.

- **Lions have the most stable personalities, and they're happiest with their lives, health, and outlook.**
- **Wolves are susceptible to mood swings and addiction, and they're less happy with their lives, health, and outlook than are other chronotypes.**

Not to suggest that all Lions can breeze through life without ever seeing a therapist, and that all Wolves need to make an appointment *today*. But there are treatable conditions that Wolves are more likely to face, like addictions, depression, and personality problems. Seeing a therapist could help.

Every day in my clinical practice I see examples of **the insomnia/depression rhythm.** The two conditions go together. Dolphins' acute insomnia can be lessened with a technique called cognitive behavioral therapy for insomnia (CBT-I). And if you treat that, you go a long way toward easing their depression, too. In an Australian study,[13] 419 outpatients from a sleep disorder clinic rated their depression and then attended a series of either individual therapy sessions or group sessions about CBT for insomnia. **At the end of the series of sessions, the insomniacs showed significant improvement in both their sleep patterns and depression symptoms.** Dolphins are not anti-therapy. In fact, with their neurotic tendencies, I'd say that Dolphins are the chronotype most likely to seek help for a problem. To any Dolphin on the fence about whether or not to see a sleep specialist, my advice is to go for one session. Just one. You don't have to commit to anything. You can find specialists at www.sleepcenters.org.

Bears are the most affable and easygoing of all the chronotypes, but in terms of **the emotional intelligence rhythm,** they could use some help. In a Spanish study[14] of over 1,000 healthy men and women between eighteen and fifty, researchers first determined subjects' chronotypes and then gave them a psych test that measured three dimensions of emotional intelligence: emotional attention (listening to others), emotional clarity (knowing what's really going on), and emotional repair (being able to fix problems). The female subjects, as a whole, were better listeners (no big surprise), and men were apt to try to fix things (naturally). As for chronotype distinctions, Bears and Wolves were better listeners than Lions. But otherwise, Bears rated the lowest on clarity and repair. It's probably because Bears like to keep things on an even keel, so they aren't inclined to look too hard at their emotions or to fix something that may or may not be broken. In my experience, Bears do tend to resist counseling, which, if they are dragged into it, slows progress.

If you decide to give therapy a try, the next step is to choose a therapist. What is **the compatibility rhythm** for selecting the right doctor for you? I'd advise you to ask a potential therapist about his or her energy and alertness fluctuations, and to choose someone who shares your rhythms. Dolphins and Wolves should go with therapists who are more alert in the evening; Lions and Bears can go with someone who is more alert in the morning and afternoon. Compatibility with a therapist is multidimensional, of course. But, from the outset, you don't want to sit across from a mental health care provider who is low-energy when you are up, and vice versa.

As far as **the scheduling rhythm** is concerned, see your therapist when you (and the doctor) are alert. Remember, your alertness peak times are when your brain is best equipped to handle strategic and analytical problems. In therapy, you need to be analytical, in part, to ensure what we call "containment," or not letting a session get so emotional that it stops the therapeutic process or that you take those raw emotions with you when you leave. I know, in movie scenes, therapeutic breakthroughs come when a patient has a sudden revelation in session and then starts crying, acting out, and so forth. In real life, therapy is more helpful when it's an intellectual process. (It's called "emotional intelligence" for a reason.) Later on, during off-peak times, you can reflect on the conversation and make insightful connections and be as emotional about what you've figured out as you like. Share the revelations you have during off-peak times with your therapist during on-peak times.

The Rhythm Recap

The life satisfaction rhythm: When you are happy about your life and future.

The insomnia/depression rhythm: When not sleeping well goes hand in hand with not feeling well, and vice versa.

The emotional intelligence rhythm: When chronotypes are able and inclined to listen, know what's really going on, and fix emotional problems.

The compatibility rhythm: When you ask a therapist about his or her chronotype, and choose someone whose type matches your own.

The scheduling rhythm: When to schedule sessions during on-peak times.

THE WORST TIME TO SEE A THERAPIST

Off-peak alertness times of the day. To keep from falling asleep during the session, and to be sure you're able to concentrate on the issues, avoid seeing a therapist in the early morning (except Lions) or mid-afternoon. And no one should see a therapist in the evening except for a Wolf.

THE BEST TIME TO SEE A THERAPIST

I heartily recommend that you seek professional help for an emotional problem, be it short- or long-term, or for expert advice about curing insomnia. To find a traditional therapist, go to www.psychologytoday .com. To find a sleep therapist, go to www.sleepcenters.org.

Dolphin: 4:00 p.m. to 6:00 p.m.

Lion: 7:00 a.m. to 12:00 p.m.

Bear: 10:00 a.m. to 2:00 p.m.

Wolf: 5:00 p.m. to 8:00 p.m.

MEDITATION

Meditation is a psychological practice anyone can do anytime, anywhere. I do it myself every morning in the shower—to put my mind in the here and now and to center myself for the day ahead. Different chronotypes can try meditation or deep breathing exercises (go to www.thepowerofwhen.com for instructional videos) at different times to quiet nerves, reduce stress, boost creativity, or clear the mind during daily transitions.

- **Lions:** Morning meditation will set the tone for the day ahead.
- **Dolphins:** Mini-meditations of two or three minutes can reduce stress on an as-needed basis. Pre-bedtime meditation can reduce cortisol levels and lower blood pressure and heart rate.
- **Bears:** Lunchtime meditation can help boost creativity in the afternoon. Evening meditation can ease the transition from work to home.
- **Wolf:** Mini-meditations of two or three minutes can reduce stress on an as-needed basis. Pre-bedtime meditation helps you transition from peak to sleep.

TAKE A BATH OR SHOWER

Failure: Bathing at the wrong time, and accidentally making yourself sleepy when you should be alert, and alert when you should be sleepy.

Success: Bathing to wake yourself up or relax yourself at appropriate times, and to incubate aha! moments.

The Simple Science

To shower in the morning or in the evening? Many of my patients have asked me this question, sensing intuitively that showering or bathing does have something to do with waking up and/or inducing sleepiness.

Picture a woman lounging in a bubble bath. This image is a visual cliché for "soothing!" Picture a man taking a shower while smelling green soap. It's a visual cliché for "invigorating!" These secondary benefits of taking showers and baths—relax or refresh—seem to be more important than the primary objective of cleaning the body of dirt and oil. In fact, **the timing of when you immerse or rinse yourself with water does affect your bio-time, and can be as "soothing" or "invigorating" as advertisements would have you believe.**

Core body temperature changes throughout the day, dropping to its lowest point around 4:00 a.m. and rising to a high point around 10:00 p.m. (for Bears). The changes in temperature affect your alertness, muscle strength, and flexibility, as well as your sleep/wake cycle. In mammals, fluctuation of two or three degrees influences the dozens of inner clocks throughout the body, according to researchers from Northwestern University.[15] Showers and baths can be tools used to serve a specific purpose: giving circadian temperature fluctuation a boost. Call it **the temperature rhythm.**

When you wake up, your core body temperature is rising. To help it rise, take a cool (not cold) shower. Lower-temperature water on the surface of your skin causes blood to flow into your core to keep your vital organs warm. So a cold shower actually makes your core temperature rise, waking you up.

When your core temperature drops at night, it triggers the release of melatonin and signals to the brain, "Time to get sleepy." A hot-water immersion will cause blood to flow to the extremities (notice how your skin gets flushed after a Jacuzzi or a sauna?) and away from your vital organs, lowering your core body temperature and encouraging sleepiness.

What about post-workout? Exercise increases "surface body temperature," or the temperature of the skin and muscles, not core body temperature. A typical quickie post-workout rinse will cool your surface temperature but won't have much of an impact on your core.

When to shower? The answer is: Whenever you shower, adjust the water temperature accordingly by time of day.

- **In the morning, take a tepid or cool shower. A hot shower in the morning will prolong sleep inertia.**

- **Before bed, a hot bath or shower will encourage sleep. A tepid or cool shower at night might suppress it.**

When will taking a shower bring on flashes of brilliance? We've all experienced something like the following: You just can't come up with the solution to a vexing problem, despite trying really hard to do so. But then you step into the shower, and the answer pops into your head when you weren't even thinking about it. Having aha! moments in the shower is a bit of a cliché, but it's quite real. **The illumination rhythm,** when you're suddenly struck with insight, is a proven scientific phenomenon.

In 1926, British psychologist Graham Wallas broke the creative process down into four distinct stages:

- **Preparation.** The groundwork of creativity, when you research or outline broad-stroke concepts or ideas.
- **Incubation.** The strange and mysterious period when you *don't* consciously think about the idea, and let your subconscious sort it out for you.

- **Illumination.** The sudden insight of creativity, when you get a flash of brilliance — the lightbulb moment.
- **Verification.** The refinement period, when you confirm that your brilliant insight actually makes sense.

The incubation period can be long — a night's sleep, a day, a week, a year. Or it can be short. Even breaks of five minutes — the length of time between stepping in the bath and . . . "eureka!" — can bring about the illumination you seek, according to numerous studies[16] on the subject.

Why does illumination happen so often in the shower? You are alone (most likely), physically distracted, mentally unfocused. You're not talking to someone, or engaged by the TV, the computer, your family, your work, a book, dinner, the gutters that need cleaning, the dog that needs feeding. If you had a waterproof TV in the shower, those eureka moments would stop happening.

Since incubation is purposely *not* thinking about something, planning a shower with the goal of illumination might defeat the purpose. But if you take a shower when your mind is primed to wander — during off-peak alertness times — you're more likely to spike your illumination rhythm.

One last personal note: I use the shower for my daily **meditation rhythm** before my day starts. I concentrate on my breathing and clear my mind while the water flows directly over my head. It is tough to do, but is *very* effective at bringing my focus to the here and now. It's my own little bio-time trick. You can try it for sixty seconds whenever you shower.

The Rhythm Recap

The temperature rhythm: When to shower or bathe to help you wake up or feel sleepy.

The illumination rhythm: When to shower or bathe for sudden moments of insight.

The meditation rhythm: When to shower as a meditation tool to clear your mind and bring you back to the present moment.

THE WORST TIME TO TAKE A BATH OR SHOWER

11:00 a.m. For all chronotypes, a hot shower mid-morning will make you sleepy when you should feel wide-awake. It's unlikely that showering during on-peak mental alertness times will bring about a creative breakthrough. Your brain is just too focused for lightbulb moments to occur at this hour.

THE BEST TIME TO TAKE A BATH OR SHOWER

Dolphin: 7:30 a.m. cool shower to help you wake up; **9:00 p.m.** hot bath to calm you down before bed.

Lion: 6:00 a.m. if you need a morning rinse; **6:00 p.m.** post-workout cool shower to stave off evening sleepiness.

Bear: 7:30 a.m. cool shower to help wake you up; **10:00 p.m.** hot bath to calm you down before bed.

Wolf: 11:00 p.m. hot bath to calm you down before bed. No morning shower. It's better for you to take that time getting as much sleep in the morning as possible.

"NOW I FEEL WEIRD IF I DON'T DO IT"

When I suggested to **Stephanie, the Dolphin,** that she take a cool shower in the morning *and* a hot bath at night, she said, "That's a lot of time in water. I'm not an actual dolphin!" After a couple of weeks of doing it, however, she said, "I get it now. The shower is the centerpiece of the new morning ritual—exercise, shower, eat—to clear my head. And the bath is the centerpiece of the new nighttime ritual—sex, bath, read—to calm me down. I thought I didn't have time for both, but they're serving a purpose and I *love* having a concrete plan. Now I feel weird if I *don't* do it." Weird how? "Foggy in the morning, and keyed up at night. That used to be how I lived every day. My old normal is the new weird."

TAKE A PILL

Failure: Taking medication when it's least effective.

Success: Taking medication when it'll do the most good.

The Simple Science

The father of chronobiology was Franz Halberg, MD, a Romanian national who spent most of his career at the University of Minnesota, where he founded the Halberg Chronobiology Center. Halberg is credited with coining the term "circadian rhythm." While treating soldiers in an army hospital during World War II, he studied daily rhythms of infection and healing. Later, at Harvard Medical School, he began experimenting on mice and discovered daily fluctuations in their body temperature, immune functioning, and ability to metabolize toxins. In a 1959 experiment,[17] he divided mice into subgroups and injected them with a poisonous ethanol solution in four-hour intervals over a twenty-four-hour period. In one group, half of the mice died. In another group, injected twelve hours later, only 30 percent died. The 20 percent survival rate difference depended solely on the time of injection. If more mice died according to the time of day they were poisoned, it stood to reason that more mice (and humans) might live according to the time of day they were treated with healing medications. In his long career (Halberg died in 2013, at ninety-three), he wrote over three thousand papers on chronobiology, conducted experiments on cancer patients in India, and studied blood pressure in patients in Japan. He oversaw the timing of his own wife's chemotherapy, and she lived longer than expected on his revised schedule.

The man was a genius, way ahead of his time. It's been over fifty years since his mice/ethanol experiment, and most physicians are still using the old "take pill once per day" prescription model. In some corners of the scientific and medical communities, the timing of medications is being rethought and reevaluated. We know that certain drugs are more effective if taken at certain times, including pills used by millions of people every day.

- **Aspirin.** In a 2014 Dutch study,[18] researchers at the Leiden University Medical Center divided 300 heart attack survivors into two groups. The first group took 100 mg of aspirin at 8:00 a.m., and the second group took the same dosage at 11:00 p.m. Blood platelets, which cause clotting and, therefore, heart attacks and strokes, are most active in the morning; aspirin has long been known to reduce platelet activity. Would the morning-dose group be in better shape after two three-month testing periods than the evening group? Surprisingly, **the evening-dose group showed greater reduction in platelet activity and tolerated aspirin-related stomach upset better than the morning group.** Why? Dosing at night prevented platelets from forming, and the patients were asleep and unaware of stomach upset.

- **Statins.** A British study[19] by researchers at the University of Sunderland set out to determine whether taking statins in the morning or evening did a better job of lowering cholesterol in the blood. Over a period of eight weeks, fifty-seven subjects took their prescribed dose at their randomly assigned time and were then tested for blood lipids. The morning group showed significant increases in total cholesterol and LDL (bad cholesterol), leading researchers to conclude that, since cholesterol is produced in larger quantities overnight, **statins are more effective if taken before bed, when they can actively fight lipids (fat in the blood).**

- **Blood pressure medication.** As you know by now, blood pressure fluctuates on bio-time, rising in the morning and dipping between 10 and 20 percent overnight. People with high blood pressure, however, don't experience a dip at night. Their pressure stays high. Researchers at the University of Vigo in Spain set out to determine the impact of non-dipping on one's risk of heart attack and stroke. The study[20] extended over five years and involved over 3,000 men and women with high blood pressure. **Nighttime medicine takers had a 33 percent lowered risk of heart attack and stroke compared with the morning pill takers.**

The dosing rhythm, the time of day when a drug is most effective, is a big unknown to most patients — and most doctors. I often ask patients, "When do you take your medicine?" The answer is usually, "Once a day" or "Twice a day," with no consideration of specific times (morning, afternoon, or night). I've noticed that, if it's more convenient for them to take all of their drugs at the same time of day—usually, in the morning—that's when they do it. And, by and large, prescribing physicians are okay with that.

It's up to doctors and pharmacists to pay more attention to the ever-increasing canon of scientific research proving that when you take a pill matters. I recommend that you ask your doctors for more information about the timing of medications. Who knows? You might wind up teaching them a thing or two and opening their minds to new, proven, successful strategies and methods of treatment.

Here are a few studies to mention to your doctor (easily found on Google) to get the conversation started:

- "Effect of Aspirin Intake at Bedtime Versus on Awakening on Circadian Rhythm of Platelet Reactivity: A Randomized Cross-Over Trial," by Tobias Bonten et al.
- "Taking Simvastatin in the Morning Compared with in the Evening: Randomised Controlled Trial," by Alan Wallace, David Chinn, and Greg Rubin.
- "Sleep-Time Blood Pressure: Prognostic Value and Relevance as a Therapeutic Target for Cardiovascular Risk Reduction," by R. C. Hermida, D. E. Ayala, et al.

The Rhythm Recap

The dosing rhythm: When to take a pill for greatest efficiency.

THE WORST TIME TO TAKE A PILL

Whenever. It's crucial to patients to take their medications mindfully and not just whenever they remember to or when it's convenient. Timing matters. Before changing your existing routine, talk to your doctor about the time at which it might be best to take your medication.

THE BEST TIME TO TAKE A PILL

People take their medication at various times, based on habit: "with meals to avoid stomach upset," at bedtime or first thing so they remember to dose, and so forth. Sometimes, those habitual or logical dosing routines are, coincidentally, on good bio-time. In other cases, doing what you've always done isn't helping you.

In the chart below, I specify general times—"before breakfast" or "bedtime"—for taking various medications instead of breaking down dose timing for each chronotype and each medication. By now, you know your bedtimes and mealtimes by heart. Plan accordingly. **As always, consult your doctor before you make any changes to your existing medication routine. This is not medical advice, but more a conversation-starter for your next appointment.**

Drug	Bio-Time Dosing
Antihistamines	evening
Aspirin	bedtime
ACE inhibitors and ARBs	bedtime
Acid reflux drugs	before breakfast
Beta-blockers	bedtime
Corticosteroids	afternoon, to reduce overnight inflammation
Heartburn pills	after dinner
Multivitamin	after breakfast
NSAIDs	four hours before maximum pain
Osteoporosis drugs	an hour before breakfast
Probiotics	with breakfast
Rheumatoid arthritis drugs	bedtime
Statins	bedtime

DANGEROUS TIMES

The reason drugs are more or less effective at particular times is because your ailments and conditions fluctuate on circadian rhythms of their own.

- **Allergies** are at their worst in the morning, thanks to nighttime accumulation of pollen. Sufferers wake up sneezing their heads off.

- **Arthritis.** Joints are stiffest between 8:00 a.m. and 11:00 a.m. because your immune system goes into overdrive at night, increasing inflammation.
- **Asthma attacks** strike most often between 4:00 a.m. and 6:00 a.m., when lung function is at its worst.
- **Depression** is at its worst upon waking at 8:00 a.m.
- **Heart attacks** are most common between 6:00 a.m. and noon due to overactive platelets and clotting proteins.
- **Heartburn** strikes after dinner when stomach acids peak and are exacerbated by lying down on the couch or in bed.
- **High blood pressure** will be at its highest at 9:00 p.m.
- **High blood sugar** will be particularly elevated when the liver dumps glucose into your system from 4:00 a.m. to 6:00 a.m. to wake you up.
- **Hot flashes** in menopausal women are most intense and frequent in the late evening around 9:00 p.m. and continuing overnight.
- **Migraines** start forming in your brain at 4:00 a.m. and are most commonly felt upon waking.
- **Restless leg syndrome** is at its most annoying at midnight.
- **Seizures** are most common between 3:00 p.m. and 7:00 p.m.
- **Strokes** are most likely to occur between 6:00 a.m. and noon.
- **Tension headaches** predictably strike in the late afternoon.

WEIGH YOURSELF

Failure: Not using a scale or using it incorrectly.

Success: Using the scale as a tool to help you lose weight and maintain a healthy weight.

The Simple Science

For people who are trying to lose weight, the scale is a scary monster lurking in the bathroom. You might be afraid to step on it. If you do hop on, seeing a higher than expected number can throw you into an emotional tailspin and set off a self-destructive eating binge.

But if you weigh yourself on bio-time, you can turn the scary monster into a simple tool that you view dispassionately, like a thermometer, to build a healthy attitude about your weight.

The low-point rhythm, or when your weight is the lowest it'll be all day, is easy to calculate for any chronotype — right after you wake up, after you void your bladder, and before you eat or drink anything. Weight can fluctuate three to five pounds in a single day, depending on what you consume (foods with a lot of salt can make you retain water; alcohol will make you bloat as well), whether you have a bowel movement, and whether you exercised. A sweaty workout might make you lose water weight. You might be tempted to weigh yourself afterward, but resist. That number is an artificial low. As soon as you rehydrate, you'll put back on the water you lost. Weight loss experts suggest that you weigh yourself at the same time every day for three consecutive days, add the numbers together and divide by three for your average weight of the week. Consistency is crucial for keeping perfect bio-time with everything else, including weight assessment. But I disagree with weight loss experts about only stepping on the scale three times per week. Weighing yourself daily is better, and here's why.

For many people, weighing themselves sets off a flood of negative self-talk and emotions. The number on the scale becomes the measure of their worth. Rationally, they know that is not true. But weight can be an emotional powder keg. If you use **the habituation rhythm,** or weighing yourself daily to get used to seeing the number, the number itself will become just a data point, not a measure of your worth. Repetition will allow you to detach emotionally, and then you can use the scale as a tool to see how your eating and exercise routines affect your weight.

The weight loss rhythm is to hop on the scale every morning on an empty stomach, naked, and to log the number in a notebook, app, or chart. A two-year University of Minnesota/Cornell University study[21] of 162 overweight adults tested the idea that daily self-weighing and tracking alone could result in weight loss. The self-weighers did lose weight—about 5 percent of their body weight—during year one. More importantly, they kept it off for year two. Lead researcher David Levitsky, PhD, told the *Cornell Chronicle* that his method "[f]orces you to be aware of the connection between your eating and your weight. It used to be taught that you shouldn't weigh yourself daily, and this is just the reverse. We think the scale also acts as a priming mechanism, making you conscious of food and enabling you to make choices that are consistent with your weight."

The keep-it-off rhythm. A study[22] by experts at Brown University divided 314 subjects who'd lost up to 20 percent of their weight in the last year into three groups: the control group, the Internet support group, and the face-to-face support group. Regular online or in-person support meetings with the latter two groups were conducted a few times over a year and a half. Of the control group, 72 percent gained back five or more pounds. Among the online support group, 55 percent gained weight back as well. Only 46 percent of the face-to-face support group gained weight back. The subjects who weighed in daily and told their support people (either online or in person) how they were doing had an 82 percent reduced risk of re-gain. Lead researcher Rena Wing wrote, "Daily self-weighing was strongly associated with successful weight-loss maintenance" and was especially effective in combination with human intervention to keep subjects honest with themselves.

The Rhythm Recap

The low-point rhythm: When your weight is the lowest (first thing in the morning, after urinating, before eating or drinking).

The habituation rhythm: When to weigh yourself so that you'll get so used to seeing the number that it won't trigger an upsetting emotional reaction.

The weight loss rhythm: When to measure and track your weight in order to lose some of it.

The keep-it-off rhythm: When to weigh yourself and seek intervention face-to-face to prevent re-gain after weight loss.

THE WORST TIME TO WEIGH YOURSELF

10:00 p.m. Your weight will skew higher if you have a full belly and a glass or two of wine. Accumulated water weight after hours spent sitting (at your desk, in the car, at the table) can account for three pounds over the course of the day.

THE BEST TIME TO WEIGH YOURSELF

For all chronotypes, step on the scale first thing in the morning, after urinating, before eating. If you are trying to take off a few, weigh yourself at the same time for an accurate measure *every single day*, and track the number in notebook or app.

Dolphin: Daily at 6:30 a.m.

Lion: Daily at 5:30 a.m.

Bear: Daily at 7:00 a.m.

Wolf: Daily at 7:30 a.m.

THE FRESHMAN FIFTEEN

Guess which chronotype packs on more pounds when they go off to college? Hint: Not Lions.

Lion Envy strikes again. In a 2013 study[23] at Drexel University, in Philadelphia, researchers collected the weights of 137 college freshmen (seventy-nine women and eighty men) and then divided the students into three groups: morning types, neutral types, and evening types. The subjects were interviewed about their eating, sleeping, and exercise habits. All of the students' BMIs were basically the same at baseline, as were other variables—workload, junk food consumption, alcohol consumption, physical activity, and sleep quality.

Eight weeks later, the subjects were weighed again. The Wolf students had gained an average of two and a third pounds. The Lions and Bears? Their BMI and weight were stable. You'd assume that the Wolves were partying harder, or calling in more dorm room pizza deliveries, but no. They weren't eating or drinking or exercising differently than the other groups. The only factor that seemed to make a difference in their weight? Chronotype. You can almost hear the evening-oriented students marching in protest with signs that read, "Biology Unfair to Wolves!"

I suspect that the key factor at play here is chrono-misalignment, or social jet lag. Freshmen with new autonomy about sleep times might be staying up late and sleeping in on some days, and on others forcing themselves to rise early for a class. Extreme sleep schedule disruptions throw metabolic hormones out of whack and prime the body for weight gain. Lions who rise and rest consistently can consume just as much pizza and beer as their Wolf classmates and not gain weight.

Sleep

WAKE UP

Failure: Waking up groggy, not being able to shake off the mental fog for hours.

Success: Waking up refreshed and alert.

The Simple Science

A reminder about what sleep inertia is: that groggy feeling when you wake up. It's been compared to being inebriated or hungover. In the world of sleep medicine, we call it "sleep drunkenness." Many of my patients describe their morning daze as "waking up stupid," as if they'd lost a few IQ points overnight. In fact, you *are* cognitively impaired when you are experiencing sleep inertia. In a 2006 study,[1] researchers at the University of Colorado at Boulder monitored nine subjects in a sleep lab and tested their cognitive powers immediately upon waking—not giving them a chance to slowly come out of their morning fog by showering, having breakfast or coffee, or exercising. The subjects were asked to do double-digit addition problems. Their morning scores were significantly lower than their scores achieved on similar tests administered later in the day. What's more, the subjects were given the same type of test after staying awake for twenty-six hours straight. The up-all-night results were better than those achieved immediately upon waking after eight hours of continuous sleep.

Sleep inertia knocks out the prefrontal cortex—the working memory part of the brain. Like a computer, it needs booting-up time. Before it boots up, your memory, concentration, decision making, cognition, and performance are impaired. Your motor skills are especially poor, in more ways than one. You won't be as dexterous with tools, and you are at greater risk for getting in a car accident. Driving while groggy accounts for 20 percent of all car accidents, or some 1.2 million per year, according to AAA Foundation for Traffic Safety statistics.

Depending on your chronotype, sleep inertia might last five to ten minutes (talking to you, Lions) or two to four hours (sorry, Dolphins).

For anyone who needs to wake up and fire on all cylinders right away—emergency room doctors and nurses, soldiers, new parents—sleep inertia is no joke. They can be a danger to themselves and others.

If only there were a way to cut through that fog so you didn't lose minutes or hours to shaking clear the cobwebs of your brain...

Of course, there are several time-wise tricks to shorten or lessen the severity of sleep inertia and help you wake up with more clarity and alertness, starting with **the sleep cycle rhythm.** The severity of your grogginess is determined by what sleep stage you're in when the alarm goes off.

Over the course of the night, you will cycle through four stages of sleep four or five times. Each complete cycle is ninety minutes long, with varying amounts of time spent in each stage, depending upon the time of night. The stages are:

- **Stage 1:** The transition between wakefulness and sleep; muscles relax and breathing slows. If you wake up during this stage, you may not think that you were asleep. This makes up about 2 or 3 percent of the night and usually occurs at the very beginning or after a brief awakening.
- **Stage 2:** Deepening sleep, when brain waves get progressively slower, with the occasional fast wave. Your body temperature decreases, your heart rate slows, and your muscles relax. This makes up about 50 percent of sleep and is the easiest stage of sleep to wake from.
- **Stages 3 and 4:** The deepest sleep, when your brain waves are long and slow (delta waves). There's no eye movement during delta wave sleep. Your blood pressure drops and respiration slows, growth hormone is released, and tissue is repaired. Your body temperature is at its lowest. This stage makes up about 20 percent of sleep and is very difficult to wake from. Most Stage 3 and Stage 4 sleep occurs in the first third of the night.
- **REM:** During REM sleep, which is lighter than sleep in Stages 3 and 4, your eyes move back and forth; heart rate, blood pressure, and body temperate rise; and muscles are paralyzed. Brain waves speed up again, becoming similar in length and amplitude to those measured when a subject is awake. This is where most dreaming occurs. REM

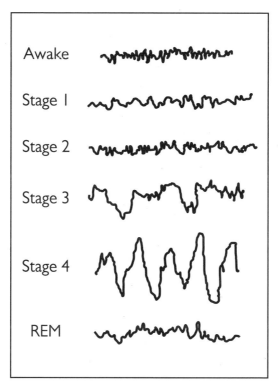

A chart of brain wave length and amplitude during the stages of sleep, from the National Institutes of Health

sleep makes up about 25 percent of sleep (if caffeine and medications do not decrease it) and is hard to wake from. You have most of your REM sleep in the final third of the night.

Most of us have to wake up at a certain time to go to work and/or get kids to school—hence the need for an alarm clock. If your alarm goes off during Stage 1 or Stage 2, sleep inertia will be minor. If it goes off during REM sleep, you'll remember your dreams, and sleep inertia won't be too bad. But if you are awakened during Stage 3 or Stage 4, it'll feel like you've been hit over the head with a frying pan.

To prevent this from happening, you have a few options. I recommend that you use a sleep monitor to track your pulse. The monitor determines sleep stage and wakes you up within a thirty-minute window of Stage 1, Stage 2, or REM sleep. For a review of products that can help with this, and my recommendations, go to www.thepowerofwhen.com.

Another strategy is not to use curtains or blinds in your bedroom. **The sunlight rhythm** simply allows you to come to consciousness as nature intended, according to the rising of the sun. That's how I do it. As dawn breaks and my room gradually brightens, my body shifts out of deep sleep into light sleep, and then into full consciousness, and I wake with minimal grogginess.

Insomnia patients are often extremely sensitive to light when they are trying to fall asleep, and for them I often recommend blackout curtains, to make the room as dark as possible at nighttime. The downside is that the room is also pitch-black when they wake up, which will exacerbate sleep inertia. Solutions: Look into buying a dawn simulator alarm clock that slowly lights up the room at the right times. Or throw back the curtains and expose yourself to direct sunlight for five to fifteen minutes immediately after waking. If you live in an area or country (far northern hemisphere or far southern hemisphere regions) where you can't get direct sunlight in the morning during certain months of the year, consider buying a light therapy box (LTB). They come in all shapes and sizes, but most use LED lights that deliver 10,000 lux white wavelength (lux is a measure of brightness) and filter out ultraviolet light. You can purchase a decent LTB on Amazon.com for less than $100. For a review of dawn simulators and LTBs and my recommendations, go to www.thepowerof when.com.

The adrenal rhythm, or getting your adrenal hormones flowing, is also an effective way to cut through the fog. Cortisol and adrenaline levels rise naturally as you wake up. Along with insulin, these hormones are the juice in the phrase "get your juices flowing." You can help to boost them by gently "stressing" your system: do a few push-ups and sit-ups, run in place, or take a cold shower.

You're probably wondering: What about coffee?

Caffeine is the most abused substance on the planet, and, despite our being conditioned to associate morning grog with a morning joe, coffee does not combat sleep inertia. Although caffeine is a stimulant, and it will make you jittery, it does not actually make you more awake. In fact, it only makes you less sleepy. Adenosine is a substance in our brains that makes us feel sleepy. Caffeine is an adenosine receptor inhibitor, effectively putting the brakes on sleepiness. This mechanism can be a wonderful thing at 2:00 p.m. Adenosine, a by-product of cellular metabolism, gradually builds up over the course of the day. But upon waking, thanks to your inner clock, you don't have adenosine in your system. Drinking coffee in the morning is like throwing water on an already extinguished fire. Coffee does give you a hit of adrenaline, which is why people equate it with feeling energized. The problem is, caffeine tolerance builds up over time, and you need to increase the amount of intake to get the effect, just as with any other addicting drug. (See "Have Coffee," page 205, to find out the best time to fill your mug.)

Instead of depending on a substance that has dubious effects, use sunlight and adrenal boosting strategies. Taking a short walk outside in sunlight whenever possible will do more to banish sleep inertia than an entire pot of coffee.

The Rhythm Recap

The sleep cycle rhythm: When to adjust waking up to the stages of sleep for an easier transition to wakefulness.

The sunlight rhythm: When to expose yourself to direct sunshine or a light therapy box to wake up your still sleepy prefrontal cortex.

The adrenal rhythm: When to jump-start cortisol and adrenaline production to help wake up.

THE WORST TIME TO WAKE UP

During deep delta wave sleep. If your alarm goes off during Stage 3 or Stage 4 sleep, it will take far longer to shake off sleep inertia.

THE BEST TIME TO WAKE UP

Dolphin: 6:30 a.m. Use a dawn simulator alarm clock to slowly illuminate the room. Get out of bed immediately and expose yourself to direct sunlight—ideally, while exercising for five minutes. Do not drink coffee until 9:30 a.m., if at all.

Lion: 5:30 to 6:00 a.m. Sleep with the curtains or blinds open or use a sleep monitor to nudge you awake gently during Stage 1, Stage 2, or REM sleep. Give yourself ten minutes before you do anything that requires brainpower.

Bear: 7:00 a.m. Sleep with the curtains or blinds open or use a sleep monitor to nudge you awake gently during Stage 1, Stage 2, or REM sleep. Do not hit the snooze button. Get out of bed immediately and expose yourself to direct sunlight—ideally, while moving for five minutes. Do not drink coffee until 10:00 a.m.

Wolf: 7:30 a.m. Use a dawn simulator alarm clock to slowly illuminate the room. Hit the snooze button once. Then get out of bed immediately and expose yourself to direct sunlight—ideally, while moving. Do not drink coffee until 11:00 a.m.

NAP

Failure: Crashing for too long, at the wrong time, and coming out of it even more tired than you were before; or not napping at all, forgoing the cognitive and creative benefits.

Success: Snoozing smart, for the right length of time and at the right time of day, to feel refreshed, recharged, and ready to perform and create.

The Simple Science

Most of us were designed by evolution to sleep for seven or eight hours a night, and to nap briefly at midday. The afternoon energy dip happens for a reason: It's your body telling you to shut down all systems for a periodic recharging of your batteries. Biologically speaking, we see a small drop in everyone's core body temperature, which releases melatonin (the key that starts the engine for sleep) between 1:00 p.m. and 3:00 p.m., depending on chronotype. In Latin America, naps are socially acceptable (ever heard of a *siesta?*). Our hardworking, hard-playing culture frowns on

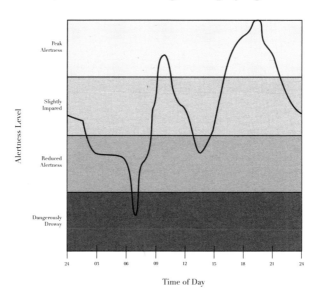

Alertness fluctuates throughout the day, peaking for Bears in mid-morning and early evening.

taking consciousness breaks—to our detriment. If we grab forty winks, we work, think, and feel better.

The performance rhythm, or the ups and downs of your ability to concentrate and get things done, gets a serious upswing from an afternoon nap. This is your brain on doze: Researchers at the University of California at San Diego proved the performance enhancements of napping objectively in their 2009 study[2] by measuring the activity in subjects' brains with an fMRI machine while they were being quizzed pre- and post-nap. Compared to the control group, the nappers' neurons were significantly brighter post-nap. In fact, nappers' brains functioned as well in the afternoon as in the morning. The non-napper group's brain activity deteriorated steadily over the course of the day.

Performance enhancement from napping is time-sensitive. Go under for too long (or too short) a time, and the effects are mitigated. Twenty-four healthy good-sleeper subjects in an Australian study[3] were divided into five groups—the no-nap control and groups that napped for five, ten, twenty, and thirty minutes, respectively—and monitored in a lab during a 3:00 p.m. nap. Three hours later, researchers asked the subjects to rate their subjective alertness, fatigue, energy, and cognition upon waking and thereafter.

- **The five-minute nappers** reported essentially the same results as the no-nappers—that is, they derived no significant benefits from the super-short nap.
- **The ten-minute nappers** were immediately and dramatically improved across the board, and the benefits were maintained for two and a half hours.
- **The twenty-minute nappers** showed delayed minor improvement. The benefits didn't kick in for thirty-five minutes, and then they lasted for two hours.
- **The thirty-minute nappers** were immediately and dramatically *impaired* in all measures for fifty minutes post-nap. When the impressive benefits finally kicked in, they lasted for another hour and a half.

Who wins napping? In terms of performance benefits and bang for your nap, the ten-minute group wins. After a short span of unconsciousness, they transitioned right into alertness, with the brain benefits lasting for the remainder of the workday.

If you nap for too long, not only will you be impaired when you wake, you won't necessarily know that you are. A Japanese study[4] tested subjects on their performance monitoring—how well they *thought* they did on a job—by breaking them into two groups, afternoon no-nappers and hour-long nappers. Not surprisingly, the nappers reported that they struggled to do the assigned task immediately after waking up, due to sleep inertia. They also reported that they still thought they did pretty well, overestimating their performance. Depending on chronotype, the post-nap fog can last for five minutes or a couple of hours. Bears and Wolves will need as long to recover from a nap as the length of the nap itself. Lions will bounce back more quickly, but their judgment and concentration will still be affected negatively from a too-long nap.

Define "too long." You never want to wake up from a nap or a night's rest while in deep delta wave sleep—that is, unless you like wading through mental molasses. So you have two choices to consider:

1. **Nap for less than fifteen minutes**—ending the nap before you enter deep sleep—to feel more alert and energized for a few hours. OR...
2. **Nap for ninety minutes** (cycling all the way back to light wave sleep) to sharpen your focus for the rest of the day.

Have you ever woken up from a nap with a bright idea? **The creativity rhythm** is real: In 2012, researchers including Andrei Medvedev, PhD, now associate professor in the Center for Functional and Molecular Imaging at Georgetown University, had their fifteen subjects wear caps with near-infrared technology to monitor the movement of oxygenated blood in their brains. While napping, there was more back-and-forth "talk" between the subjects' right hemisphere (creativity and insight) and left hemisphere (logic and analysis) than while awake. What's more, the right hemisphere was more active within itself, too. The results suggest that your brain is housekeeping, sweeping out the cobwebs, sorting, remembering, and firing globally while you are at rest. Even an hour's sleep will light up cross-hemisphere communications and allow you to make extraordinary mental connections.

The learning rhythm—for adults, too, not just kids—or your ability to interpret new input, speeds up and goes deep while you are asleep for short periods of time. According to researchers at Harvard,[5] an hour-long nap is as good as a night's sleep for enhancing learning over the course of the day, and the benefits last for twenty-four hours.

I'm sure everyone reading this would love to go flop down on the couch and take a nap to wake up brilliant *at this very moment*. But **the logistics rhythm** might prevent you from doing so. It's a sad biotime irony that the ideal time to nap—for Bears, it's 2:00 p.m. to 3:00 p.m.—is when you have to pick up the kids from school, or when afternoon meetings are often called. Even if your calendar is clear at midday, you might be in an open-plan workspace where dropping your head on your desk would be misinterpreted, no matter how passionately you argued, "I nap in order to work harder and better! I swear!"

The proof exists. It's right here on these pages. But the culture hasn't caught up yet with the science. One day, I bet offices will routinely turn off the lights at 2:30 p.m. for fifteen minutes. They should! A little nod improves alertness, creativity, learning, pretty much every skill you need in the workplace and beyond. Until the day of enlightenment comes and nap pods are standard office equipment like chairs and computers, what can you do besides lock yourself in a bathroom stall with earplugs and an eye mask?

It's a challenge. If you can find a quiet place to rest, set your alarm for ten minutes hence and close your eyes. Since the benefits of napping last for twenty-four hours, nap on Sunday afternoons to recharge your creativity and your learning capacity for Monday. My advice is to do what you can, and don't write it off as impossible. Where there's a will, there's a nap. If you can manage just ten minutes, you'll operate at a huge advantage for the rest of the day.

Maybe locking yourself in the bathroom isn't such a bad idea after all.

The Rhythm Recap

The performance rhythm: When, and for how long, to nap for increased mental cognition.

The creativity rhythm: When, and for how long, to nap for improved creative connection-making.

The learning rhythm: When to nap to enhance your brain's ability to absorb new information.

The logistics rhythm: When to find a place and a time to take a much-needed nap.

THE WORST TIME TO NAP

7:00 p.m. A nap within four hours of bedtime will relieve the sleep pressure that's built up throughout the day and could prevent you from falling asleep at night.

THE BEST TIME TO NAP

To calculate the ideal nap time, I used the Nap Wheel, the invention of Sara Mednick, PhD, author of *Take a Nap!* and researcher on a few of the studies noted in this chapter. According to her data, the Ultimate Nap occurs approximately seven hours post-wake-up, give or take a few minutes, when you are likely to strike the perfect napping balance between slow wave sleep and REM sleep, leaving your mind refreshed with minimal grogginess upon waking.[6] If you nap before seven hours post-wake-up, your nap will be comprised heavily of REM sleep, making you more creative. If you nap after the seven-hour post-wake-up, the nap will be slow wave sleep and will be more physically restorative.

Dolphin: No naps. Napping relieves sleep pressure, the opposite of what you need. Sorry, but a nap would do you more harm than good.

Lion: For a wake time of 6:00 a.m., the Ultimate Nap time is **1:30 p.m.**

Bears: For a wake time of 7:00 a.m., the Ultimate Nap time is **2:00 p.m.**

Wolf: Naps are not ideal for Wolves if they want to fall asleep by midnight. But if you really need to refresh, for a rise time of 7:30 a.m., the Ultimate Nap time is **2:15 p.m.**

DISCO NAPS

In case you missed the seventies and eighties, a "disco nap" is a late-evening snooze before you go to a club and dance until dawn. Not that I advocate going off your chronorhythm and staying up late—I discourage irregular sleep patterns—but for the occasional special event or celebration, a disco nap will keep you up and will prevent you from dragging. Follow these time-wise tips:

1. **Shoot for a full cycle.** If possible, nap for ninety minutes. After a slightly groggy reentry into consciousness, you'll be restored, with energy to burn.

2. **Drink a cup of coffee before you take the nap.** I call this strategy "nap-a-latte." If you only have time for a quickie nap, drink coffee before putting your head down. Caffeine doesn't hit your system until twenty minutes after you drink the coffee. So have a cup pre-nap, and wake when the caffeine and adrenalin hit your bloodstream. I have advised many of my Fortune 100 clients to do this, and they tell me that it keeps them alert for up to four hours.

3. **Wake up at your usual time.** Even if you partied until dawn, start the day on schedule, even if that means you sleep for only one or two hours. It'll be rough, for sure, but you'll suffer more long-term by sleeping in and throwing your schedule completely out of whack. The choice is either one bad day or a week of chrono-misalignment and the fatigue, irritability, and Swiss cheese memory that go with it.

SLEEP IN

Failure: Staying in bed hours later than usual on the weekends, causing chrono-misalignment and its symptoms (fatigue, poor focus, irritability).

Success: Staying in bed for less than an hour on weekends, preventing chrono-misalignment and its symptoms.

The Simple Science

Bears and Wolves are all too familiar with the phenomenon of Sunday Night Insomnia. After sleeping in late on Saturday and Sunday mornings, you aren't tired when you get in bed Sunday night. Your mind starts racing as you lie there, going over everything you need to do during the workweek and then, with increasing anxiety as you watch the hours tick by, wondering if you can manage to do it on no sleep. Somehow, by sheer force of will, you survive an exhausted Monday and drag through Tuesday. By Wednesday, you're approaching bio-time, which is good. But you've accrued hours of sleep debt that your body wants to collect. So what happens? You wind up sleeping in again on Saturday morning, and the vicious cycle continues.

Don't get me started on staying up late on Friday and Saturday nights.

Let's face it, you are a social creature (some chronotypes more than others), and you crave interaction. Most of the fun stuff occurs late on Friday or Saturday night, because *we all* think we can sleep in on Saturday and Sunday morning and get away with it. Wolves, weekends are your realm. Finally, two days a week when everyone aspires to your bio-time. But there are consequences. Their severity depends on you. Personally, I like to stay up later on the weekend nights. My wife and I go to the movies, to dinner, and have fun with friends or the kids. Those late nights are *valuable* time. I just don't make the mistake of sleeping in, no matter how painful waking up may be.

When you sleep in, you throw your body out of its natural circadian rhythm and cause **chrono-misalignment rhythm.** The symptoms are:

- Tiredness
- Irritability
- Restlessness
- Poor concentration
- Sleep inertia

By some estimates, 70 percent of the population suffers from chrono-misalignment or social jet lag every week. When you fly to Paris or Hong Kong and experience actual jet lag, you'll adjust quickly because your body will take cues from the rise and fall of the sun wherever you are, and you'll get back on bio-time. On average it takes about one day per time zone crossed, but that equation can be sped up with the proper timing of melatonin, light, and caffeine.

With social jet lag, on the other hand, you throw yourself out of sync with the sun by working late, sleeping late, partying late. You can calculate your chrono-misalignment fairly easily. Just estimate the difference between your workweek alarm wake time (say, 7:00 a.m.) and your weekend no-alarm wake time (say, 9:00 a.m.). In that case, your lag is two hours. You might think, *Great! Two fewer hours of sleep debt.* But it doesn't work that way.

You created the debt in the first place by sleeping in (and staying up late). **The debt you rack up in a few days of not sleeping well and waking up exhausted can't be balanced just by sleeping in on the weekends. You'll never catch up, and you'll have to operate at a deficit habitually.**

The type of sleep debt you are creating by staying up late is not replaceable by the type of sleep you get by sleeping in. For example, if you normally go to bed at 10:30 and stay up until midnight on the weekend, you miss one sleep cycle of mainly physically restorative deep Stage 3 and Stage 4 sleep. By sleeping in the next morning, you get more REM sleep, *not* more Stage 3 and Stage 4 sleep. You end up *not* feeling physically refreshed. Your circadian rhythm stays consistent even if your social or environmental rhythm does not.

I'm sorry to report that Wolves are most likely to fall into this vicious pattern.[7] Sufferers attempt to counter the effects with stimulants like cigarettes and coffee, and by eating high-sugar, quick-energy foods. Perhaps that explains **the obesity rhythm** of sleeping in. In a 2012 study[8] by researchers (including Till Roenneberg, who coined the term "social jet lag") at the University of Munich's Institute of Medical Psychology, 65,000 subjects reported their sleep patterns throughout the week. The two-thirds who logged an hour of lag between their workday and weekend schedules were three times as likely to be overweight as those who didn't lag at all. The greater the difference between workweek and weekend schedules — 10 percent of the subjects reported a lag of three hours — the higher the subject's BMI. A 2015 British study[9] found that, of 800 subjects, the ones who lagged by two hours or more had higher BMIs and biomarkers for diabetes and inflammation than those without social jet lag who kept a consistent sleep schedule.

The Rhythm Recap

The chrono-misalignment rhythm: When sleeping in causes social jet lag and symptoms including irritability, fatigue, restlessness, and poor concentration.

The obesity rhythm: When sleeping in on the weekends results in a higher BMI or obesity.

THE WORST TIME TO SLEEP IN

Saturday and Sunday mornings. Every time you sleep in, you are creating chrono-misalignment and compromising your metabolism, appetite, cognitive powers, and energy. It's all *bad,* especially if you lie in bed for two additional hours or more.

THE BEST TIME TO SLEEP IN

Have you read this section closely?

There is no best time to sleep in!

If you stay in bed on the weekends for two hours longer than usual, you're far more likely to be grouchy, fat, and sick. If you stay in bed for *less than an extra hour* on the weekends, though, you're statistically safe from chrono-misalignment's ill effects. If you absolutely *must* sleep in, do it for an extra thirty to forty-five minutes only. Thus:

Dolphin: On mornings when you are not working, you can sleep until **7:15 a.m.**

Lion: On mornings when you are not working, you can sleep until **6:45 a.m.**

Bear: On mornings when you are not working, you can sleep until **8:00 a.m.**

Wolf: On mornings when you are not working, you can sleep until **8:15 a.m.**

REALITY CHECK

Right now, a lot of people who wouldn't dream of waking up on a Sunday at 8:00 a.m. are ready to throw this book out the window, along with their alarm clock. I love long, lazy Sundays in bed, too. You can stay in bed as long as you like, snuggling, reading, sipping a hot beverage (not coffee until 10:00 a.m.), watching a movie. Stay in there all day long, as long as you're awake. Even if you went to bed at 4:00 a.m. Sunday morning, I'd still urge you to wake up at the recommended time, and take a nap later on. Otherwise, if you sleep many hours past your workday wake time, you will suffer all week long. Do what you will, sleep in or wake up. It is your life, your choice, your schedule. Just be aware of the consequences. Is an extra hour or two of REM sleep on Sunday really worth dragging from Monday to Thursday?

"I CONKED OUT ON SUNDAY NIGHT"

"Sleeping in is one of my favorite times of the week. I work hard from Monday to Friday, and on Saturday mornings, I get to chill in bed until ten or eleven. It's a luxury, and Dr. Breus wants to take that away from me!" said **Ben, the Bear.** "Not happy about that. But I committed to following the chronorhythm for one week. So on that first Saturday, I set my alarm for 8:00 a.m., even though my wife and I stayed up until 1:30 a.m. the night before. So I got only six or so hours of sleep, felt like crap, and had to ask, 'How does this mean I'm going to get more sleep?' I did the same thing on Sunday morning, after staying up late Saturday night, too, and was resentful and tired all day. But, Sunday night, when I always deal with insomnia until two or three in the morning? I conked out at 11:00 p.m. and slept so well, I felt great on a Monday morning for the first time in a decade. The big question: Is it worth it to feel a little tired all weekend, and to feel great on Monday morning? You have to strike a balance. In the next weeks, I made a point of staying up late on only one weekend night, and that helped, too. It's a trade-off. For me, over time, I decided that feeling good on Monday morning was worth being a little sleepy on Sundays."

GO TO BED

Failure: Hitting the hay later than you should and/or at inconsistent times, resulting in inadequate rest and increasing your risk of sleep-deprivation-related diseases and conditions.

Success: Consistently turning in at an appropriate time with a large enough window to get sufficient physical restoration and mental consolidation.

The Simple Science

What is a sufficient amount of sleep? I get asked this question at least five times a day. The answer is: It depends. Babies need twelve to eighteen hours per day. Toddlers need eleven to thirteen hours. School-age kids need ten to twelve. Tweens need ten or eleven. Teens need eight to ten. When, usually in their early twenties, young adults stop growing, the need for sleep diminishes. The National Sleep Foundation recommends between seven and a half to nine hours per night for adults and seniors.

I have observed that fixating on a number causes a lot of sleep disruption—in the same way that trying to reach a "goal weight" causes a lot of yo-yo dieting, frustration, and disappointment. Sleep need is genetically based. I talked about "sleep drive" back in chapter 1. Some of us are genetically programmed to need more sleep than others. Those with a long PER3 gene are most vulnerable to the damage caused by sleep deprivation, including illness and conditions such as heart disease, stroke, diabetes, obesity, depression, and cognitive/memory problems. You might have a short PER3 gene and be able to get by on six hours a night safely. You might have a long PER3 gene and feel horrible without seven or eight hours of sleep.

So how many hours to aim for? What's the hard *number*?

First let me dispense with . . .

THE EIGHT-HOUR MYTH

Everyone's sleep number is different. Some people are fine on six hours or less, as was the case for Thomas Edison. Some can't function with less than eight. I have found that adults who sleep nine hours or more per night are depressed or narcoleptic. Excessive sleep is usually a symptom of an underlying condition. Also, quality of sleep is as important as quantity, if not more so. Eight hours of disrupted sleep isn't better than six hours of quality sleep.

You've heard the expression "strive for five" about fruit and vegetable servings per day? I say to most of my patients, "Shoot for seven hours of sleep per night."

Eight hours per night would be great, but that number isn't realistic for most people, given the demands of life. It's also not exactly how our bodies were designed for optimal functioning. In an ideal world, you'd sleep for six or seven hours per night and take a ninety-minute nap midday, bringing your cumulative sleep to seven and a half to eight hours per twenty-four-hour day.

As I'll explain in detail below, you shouldn't count sleep in terms of hours but rather in ninety-minute sleep cycles. Five complete cycles (or seven and a half hours) is plenty for all physical and mental restoration. Five incomplete cycles (just seven hours) is enough for cognitive function, appetite control, memory, and healing. Four complete cycles (six hours) is not enough for Lions, Bears, and Wolves. The research is still pending about whether insomniacs actually need as much sleep as other types. Dolphins can, and do, function on less than seven hours, and the guilt and anxiety of not getting eight hours contributes to their inability to fall sleep. So, for the insomniacs in my practice, I say shoot for six hours, and if they get a bit more, *great*.

Now, how to make sure you get sufficient hours? Maybe go to bed early?

Bad idea.

For all chronotypes, going to bed *too* early, even if you have some sleep to catch up on, will backfire. Your bio-time won't adjust itself just because you are ready to turn in early, and you'll wind up lying awake, feeling tired and wired and annoyed that you're not passing out. An intense emotional response to not sleeping could keep you awake beyond your regular bedtime.

Instead, figure out your correct bio-time bedtime, and get on a **calculated rhythm.** Count backward from your regular weekday wake time by seven hours and twenty minutes (twenty minutes is the average length of time it takes to doze off). If you need to wake up at 7:00 a.m., your bedtime would be 11:40 p.m.

A slightly more complex method—the one I recommend—is to count backward from wake time by sleep cycles. Each cycle takes ninety minutes to complete. During the first two cycles of the night, you'll get the bulk of physically restorative deep delta wave Stage 3 sleep. During the last two or three cycles, you'll get the bulk of lighter, dreamy, memory-consolidating REM sleep.

To get *all* the benefits of sleep and wake refreshed, you need to complete at least four cycles, preferably five. If you wake up mid-cycle, sleep inertia will be hell.

Calculate your sleep time by counting backward from your wake time by seven and a half hours (90 minutes x 5 complete cycles = 450 minutes) plus twenty minutes (fall-asleep time), or a total of 470 total minutes. Or, wake time – 470 minutes = bedtime. Apply this simple formula to get on **the calculated rhythm for Lions and Bears:**

Lion: 6:00 a.m. – 470 minutes = **10:10 p.m.**
Bear: 7:00 a.m. – 470 minutes = **11:10 p.m.**

A formula is only as good as your ability to comply with it. Lions and Bears are likely to be able to stick with this rhythm as long as they don't commit bedtime self-sabotage, which I'll discuss in a few pages.

Dolphins and Wolves, however, will have a hard time falling asleep in the twenty-minute window allotted in the formula above, which raises the question, "If you can't fall asleep fast enough to get five complete cycles, should you just stay awake for another ninety minutes and settle for four?" The answer: not exactly, but sort of.

One of my hard rules for patients in my clinical practice is "Don't get in bed unless you are tired." The Dolphin chronorhythm is devised to help them settle down in the pre-bed hours so that they can relax. But on some nights, "tired and wired" will dominate "tired," and they won't feel the slightest bit sleepy at bedtime. If they get in bed anyway, not falling asleep right away could set off **the anxiety/insomnia rhythm** that will keep them up all night.

Instead of setting themselves up to fail at getting five complete cycles, Dolphins should shoot for four. (Actually, Dolphins might be completing five compressed cycles into the time it takes a Bear to complete five. There is a theory that insomniac sleep cycles are shorter than a full ninety minutes, and that's why they need less sleep to be functional.) Give yourself twice as much time—a full forty minutes—to fall asleep. During those forty minutes, you might use a cognitive behavioral strategy to fall asleep called Bed Restriction (see box on page 188) or try a relaxation exercise, meditation, or deep breathing. Go to www.thepowerofwhen .com for video instructions about relaxation strategies.

(Ninety minutes x four cycles) + forty minutes = 400 minutes. The formula for Dolphins is: Wake time – 400 minutes = bedtime. **Their calculated rhythm**: .

Dolphin: 6:30 a.m. – 400 minutes = **11:50 p.m.**

In their ideal world, Wolves would get in bed at 1:00 a.m. and sleep until 9:00 a.m. But the reality is, Wolves live in a Bears' world, and they have to wake up at an earlier hour to get to work and deal with family. It's a bad idea for Wolves to get in bed unless they're tired—they risk setting off **the anxiety/insomnia rhythm** and losing hours to tossing and turning. The *earliest* my Wolf patients feel sleepy is at midnight, and that's when they should get in bed. Wolves have to train themselves to fall asleep quickly. (Some of you might be thinking, "Yeah, right. Like *that's* going to happen.") It can be done by a strategy called Bed Restriction (see box). If Wolves follow my bedtime recommendation and learn to pass out fast, they are capable of getting close to five complete sleep cycles, factoring in twenty minutes of semiconsciousness between the alarm going off and rising from bed. Twenty minutes of drift time—when brain waves are similar in length and amplitude to those measured during REM sleep—has been built into the Wolf chronorhythm for this very reason.

(Forty minutes of fall-asleep time + twenty minutes of drift time) + (ninety minutes x four cycles) = 420 minutes. The Wolf formula: Wake time – 400 minutes = bedtime. **Their calculated rhythm:**

Wolf: 7:00 a.m. – 400 minutes = **12:00 a.m.**

BED RESTRICTION

The goal: Program your mind to associate "bed" with "sleep" so that when you get in it, you crash quickly.

The strategy: Bed restriction will cause increased sleep deprivation. This, in turn, causes an overflow of adenosine. The overflow helps Wolves boost their natural sleep drive and fall asleep quickly. I also use this technique with Dolphins, but they often need to be monitored more closely. In short, do not get in bed *for any reason other than sleep and sex*. No reading, thinking, TV watching in bed at all. Don't even sit on the edge of the bed while you're putting on your shoes.

Troubleshooting:

1. **"What about lying awake in bed with the lights out when I have insomnia?"** Restrict that time in bed, too. If you have been lying awake for twenty minutes, get out of bed and sit in a chair for fifteen minutes while counting your breaths or doing progressive muscle relaxation (first relaxing your toes, then your ankles, and so forth—until you reach your forehead and scalp; video instructions at www.thepowerofwhen.com). Then return to bed to try to sleep again. Repeat as needed, even if you're in and out of bed for several cycles before finally falling asleep.

2. **"How long before this works?"** A week to ten days. If you can commit to using the strategy, you will soon be falling asleep quickly with amazing regularity. In my fifteen years of practice, having worked with hundreds of patients, I can count on one hand the number of people who didn't respond to this strategy, and they had severe secondary health conditions that complicated the problem.

Even if you know when you should go to bed, that doesn't mean you will. A problem of modern times, **the procrastination rhythm,** is why so many of us stay up past our bedtime even though we know how depleted and irritable we'll feel the next day. Sleep procrastination disorder is real, and quite dangerous. Dutch researchers at Utrecht University surveyed[10] 177 subjects about their sleep behavior and found that half of them purposely delayed sleep ("one more episode," "one more link") at least twice a week. The procrastinators tended to be impulsive, with low self-control—traits associated with Wolves.

Make no mistake, procrastination is a choice—as are smoking, drinking excessively, and eating cheeseburgers instead of vegetables—and it is just as harmful to your health as those choices, if not more. Sleep deprivation increases your risk for heart disease; diabetes; stroke; obesity; depression; poor memory, concentration, and judgment; prematurely sagging skin; and premature death. Researchers at University College of London Medical School[11] kept track of the behavior and health habits of 10,000 British civil servants aged thirty-five to fifty-five (at the study's inception) over two periods of ten years. Those who once slept for seven-plus hours per night but then changed their sleep habits for whatever reason to five or fewer hours had double the risk of dying for any reason.

I often talk to patients about the importance of an established bedtime, and I try not to sound like I'm nagging. Honestly, I get it. *Homeland* is a really good show, and Candy Crush is addicting. The psychology of wanting to stay up late goes back to your parents yelling, "Go the &%$# to sleep already!" It's about rebelling against societal rules.

But bedtime is a *biological* rule. Such rules exist for one reason: to keep you alive and thriving. If you broke the biological rule of eating nutritious food and instead scarfed nothing but junk, you would get fat while starving your cells, and you would eventually succumb to diabetes, heart disease, and stroke. Breaking the biological rule of not getting enough sleep increases your risk of succumbing to the same deadly diseases.

An easy solution to keep you from procrastinating: Set your alarm for one hour *before* your bedtime, and start your Power-Down Hour (see box below)—a time to relax, unwind, and *shut down all screens.* If the TV is off, you won't be tempted to binge-watch. If the computer is off, you won't view one more cat video. If your phone is off, Words with Friends will just have to wait until tomorrow.

POWER-DOWN HOUR

During the hour before bedtime, downshift. Bring on that sleepy feeling by doing off-screen activities that lower cortisol level, body temperature, and blood pressure and boost melatonin. I ask patients to break up the hour before lights-out into three twenty-minute sections.

Section One (first twenty minutes). Do the things you must get done—if you don't, you will think about them while you're trying to fall asleep. These may include:

- Make a to-do list for the morning.
- Journal.
- Get kids' backpacks or your work materials together.
- Prepare for tomorrow in whatever way may make your morning life easier.

Section Two (second twenty minutes). Do your nightly hygiene routine—which may include taking a hot bath or shower—in a dimly lit bathroom, or one with the special sleep bulbs that filter out blue wavelength light.

Section Three (last twenty minutes). Do a relaxing activity, such as:

- Light stretching or "bed yoga."
- Read paper magazines, the newspaper, or a book (including this one!).
- Have casual conversations with friends and family. You can use your phone, but only to make and receive calls.
- Play cards or games (but do not get too excited or competitive).
- Meditate.
- Pray or read Scripture.

The Rhythm Recap

The calculated rhythm: When you should go to bed—a clock time calculated by subtracting the number of minutes needed to complete at least four sleep cycles and account for sleep onset from the hour you have to wake up.

The anxiety/insomnia rhythm: When you get stuck in a cycle of worrying that you're not getting enough sleep, which causes wakefulness.

THE WORST TIME TO GO TO BED

More than eight hours before you need to wake up. Don't try to get extra sleep by going to bed early. Stick to your set bedtime to prevent circadian disruption.

THE BEST TIME TO GO TO BED

Dolphin: As close to 11:30 as possible.

Lion: As close to 10:00 p.m. as possible.

Bear: As close to 11:00 p.m. as possible.

Wolf: As close to midnight as possible.

SLEEP PARTNER CONFLICTS

When you and your partner have different ideas about bedtime, conflicts arise, and these may have consequences. New studies of "dyadic sleep," or sleep harmony, look at how the interaction of two individuals at night may affect their sleep and their relationship generally.

Anecdotally, I've noticed that the sleep lag between people of any two different chronotypes is usually no more than an hour or two. It's unlikely that a Lion with a 10:00 p.m. bedtime and a Wolf with a 12:00 a.m. to 1:00 a.m. bedtime would meet socially, let alone fall in love and pair off. But I have met many Wolf/Dolphin pairs (they both go to bed late), as well as same-type pairs, Lion/Bear pairs, and Bear/Wolf pairs.

Obviously, a Dolphin's tossing and turning can disrupt the sleep of her partner. A Bear's REM sleep snores might wake up a Wolf, who needs every minute of sleep she can get. One partner's coming to bed later, being in bed with a device on, waking earlier, being a light sleeper—any one of these may lead to discord. Many patients have told me that they spend little time with their spouse due to conflicts in sleep/wake patterns. All of the frustration, disrupted sleep, and snores can cause emotional and health problems[12] for the individual and erode the marital bond.

The truth is, you don't need to have the same sleep schedule as your partner to be in a happy relationship. But you both need your sleep, and if the bedtime conflicts prevent you from getting it, other issues might come up that damage the bond between you. Sleep deprivation can lead to irritability, cognitive failure, and poor health.

Since you can't change your chronotype, the best advice I can give is to work as a team to help each other get as much sleep as possible. And/or you can nudge both your chronorhythms closer together. Early risers should be very quiet, not turn on the lights, and be courteous to their still-sleeping partners in the morning. The up-late partner can start using 0.5 mg of melatonin ninety minutes before bed for earlier sleepiness. If things are really bad and neither of you is sleeping well, you might need professional help (take a look at "See a Therapist," page 149), or you might need to sleep in separate rooms temporarily until both of you make up your sleep debt and you can face the problems constructively and clearheadedly.

CHAPTER 10

Eat and Drink

EAT BREAKFAST, LUNCH, AND DINNER

Failure: Eating daily calorie intake out of sync with bio-time, making you gain weight and putting you at risk for illnesses such as diabetes and heart disease.

Success: Eating daily calorie intake on bio-time, helping you lose weight, have more energy, feel less hungry, and ward off diabetes and heart disease.

The Simple Science

Until now, you've probably only concerned yourself with the "what" of food, the breakdown of macronutrients on your plate, carbs vs. protein, etc. There are so many rules and warnings. I totally understand why it's practically impossible to stay on or stick to a healthy eating regimen. One year, fat is bad. The next year, fat is good. Then certain kinds of fats are good. Or maybe some bad fats are actually okay? Even if you are up-to-date about the ever-changing rules and know more about nutrition than your doctor, it's likely weight loss is still a challenge. And it's not going to get easier as you age.

As soon as a new patient walks in the door of my office, I can tell in five seconds if he is out of sync with his bio-time. The giveaway is how much weight he carries around the midsection. **There is a direct relationship between chrono-misalignment and belly fat.** If your bio-time is in good shape, so is your body. If not, digestive problems and metabolic dysfunction are bound to happen.

As I've explained, your body has dozens of inner clocks. When they are all on time with the rise and fall of the sun and with each other, you are a perfectly running machine that does everything at the right time, in the right order, in a perfectly synchronized rhythm. Digestive and metabolic body parts that keep their own timetable:

- **The liver.** Not only is the liver the body's filtration system, it also controls glycogen (sugar), cholesterol, and bile production and distribution.

- **The pancreas** is in charge of insulin and the ebb and flow of blood sugar.

- **The gastrointestinal tract** keeps time moving food along and absorbing nutrients that every cell in our bodies needs to stay healthy and do its job.

- **Muscles.** We don't automatically think of muscles as a metabolic organ, but when your body releases fat and sugar for energy to, say, climb some stairs or blink an eye, they go to the muscles to be consumed.

- **Fat cells.** Each fat cell in your body produces hormones that tell you when you're hungry or full, among other digestive and metabolic jobs.

I haven't even begun to delve into the complexity of genetic proteins and enzymes in digestion's multistep processes. (For example, the BMAL1 gene initiates the enzyme NAMPT to produce the chemical NAD, which activates the SIRT1 gene to secrete the hormone insulin, which leads to the release of glycogen or the storage of fat... and so on, and so on, times a hundred.) If any one of your digestive or metabolic compounds or genes is out of sync, the whole system is disrupted. It's like removing a small but important component, causing the entire machine to grind to a halt.

Or, how about this metaphor: Digestive processes are like trains leaving the station. One goes through a tunnel, followed by another and another. Every train stays on schedule and gets where it's supposed to go without incident.

Now picture two, four, a hundred trains heading for the same tunnel at the same time. When the schedule is out of sync, the result is a massive pileup.

When you eat out of sync with bio-time, the result is a massive pileup of pounds. Excess fat and dysfunctional metabolic hormone functions bring on inflammation and oxidation, which causes just about every disease known to man, especially heart disease, cancers, and diabetes. This can be avoided, even reversed, by directing your dietary attention toward the "when" of eating, and, for the time being, forgetting about the "what." If you changed nothing about "what" you ate and only changed "when" you took in calories, you would lose weight.

One way is to follow an **eating time restriction rhythm.** A 2012 study[1] by researchers at the Salk Institute for Biological Studies, in La Jolla, fed one group of mice high-fat food around the clock. The other group was given the same amount of food and the same menu but could eat only within an eight-hour window. The twenty-four-hour buffet mice became obese and diabetic. The mice whose mealtimes had been restricted gained hardly any weight and remained healthy.

In a follow-up study,[2] Salk Institute scientists divided mice into four groups of eaters (high-fat, high-sugar, high-fat *and* high-sugar, and the kibble control) receiving the same number of calories per day. Within each group, the mice were divided into time-restricted eaters (feeding windows of nine, twelve, or fifteen hours) and unrestricted eaters (feeding around the clock). Predictably, the mice that ate at all hours were obese and diabetic by the end of the thirty-eight-week study. The nine- and twelve-hour-window eaters from all food content groups stayed slim and healthy.

Fascinating side note about this study: Some restricted-eater mice were allowed to eat round-the-clock on the weekends, and they didn't gain weight. Some mice were switched from 24/7 access to food and put on a restrictive eating schedule. They lost all the weight they'd previously gained. Although these are mice, not human, studies, it's fair to say that eating within an eight- or twelve-hour window makes complete sense for our kind as well, since the GI tract runs on a four-hour ultradian rhythm, smaller time loops within the twenty-four hour circadian rhythm.[3] **Eating every four hours (within an eight- or twelve-hour window) helps you keep perfect digestive bio-time.**

Studies have confirmed the importance of "when" for weight loss in humans—in particular, **the early eating rhythm.** In a Spanish study,[4] 420 overweight or obese men and women were put on a diet of 1,400 calories per day for twenty weeks. Half of the subjects were early eaters, having their biggest meal of the day before 3:00 p.m. The other half, the late eaters, had their biggest meal after 3:00 p.m. They ate the same quantities of the same food as the early eaters, exercised at a similar intensity and frequency, slept the same number of hours, and had comparable appetite hormones and gene function. Which group lost more weight? **The early eaters lost, on average, twenty-two pounds; the late eaters, seventeen pounds, a 25 percent difference.** The late eaters were more likely to skip breakfast, a huge mistake if you're trying to lose weight. In a study[5] of some 27,000 men over a sixteen-year period, researchers at Harvard School of Public Health determined that skipping breakfast increased your risk of coronary heart disease by 27 percent. Men who ate late at night had a 55 percent higher risk.

Why does eating early make human and rodent subjects lose or maintain weight and ward off metabolic diseases? That pattern is a zeitgeber—a German word that means "time giver." Zeitgebers are external forces that help your inner clocks get on perfect bio-time. The most powerful zeitgeber is the rise and fall of the sun. Another is the outside temperature. Another? An early and time-restrictive eating schedule.

When sunlight hits your eyeballs and travels along the optic nerve to jangle the cluster of neurons called the SCN, your brain's biological clock knows the day has begun. When you take your first bite in the morning and food travels down your esophagus and hits your stomach, your second brain (the gastrointestinal tract) knows the day has begun. When those two events—morning exposure to sunlight and morning intake of food—happen at roughly the same time, you brain and gut clocks will be synchronized. You will digest and convert food into energy with efficiency and, in turn, store less fat and feel more vital.

On the other hand, if you skip breakfast, your first brain might know it's daytime, but your second brain lags behind. In this conflicted state, your body doesn't know what time it is. When you get around to eating later on, your digestive mechanisms are confused and inefficient, resulting in inflammation (heart disease) and dysfunctional hormonal response to sugar/fat uptake (diabetes), resulting in stored fat and lethargy.

Here's how an ultradian-rhythm eating schedule would look for Bears:

> Eat a big breakfast within an hour of waking, say, at 8:00 a.m.
> Eat a medium-sized lunch four hours later, at 12:00 p.m.
> Eat a small snack four hours later, at 4:00 p.m.
> Eat a small dinner at 7:30 p.m. and make sure your last bite is before 8:00 p.m., to stay within the twelve-hour eating window.

In the chronorhythm for each chronotype outlined in Part One of this book, the eating times are not exactly four hours apart, because there are other important activities to get done during the day besides eating. But my schedule is pretty close to the every-four-hour ideal. My imperfect schedule has only three rules:

1. Eat within one hour of waking.
2. Eat like a king or queen at breakfast, a prince or princess at lunch, and a pauper at dinner.
3. Take your last bite three hours before bedtime.

The Rhythm Recap

The eating time restriction rhythm: When to have your breakfast, lunch, and dinner within an eight- or twelve-hour daily eating window.

The eating early rhythm: When to eat the bulk of the day's calories to prevent weight gain, heart disease, and diabetes.

THE WORST TIME TO EAT BREAKFAST, LUNCH, AND DINNER

Outside a twelve-hour window that starts and ends early.

THE BEST TIME TO EAT BREAKFAST, LUNCH, AND DINNER

The meal breakdown below does not include strategically timed snacks. See "Snack," page 214, which rounds out the daily nutritional story.

Dolphin: B, 8:00 a.m.; L, 12:00 p.m.; D, 7:30 p.m.

Lion: B, 6:00 a.m.; L, 12:00 p.m.; D, 6:00 p.m.

Bear: B, 7:30 a.m.; L, 12:30 p.m.; D, 7:30 p.m.

Wolf: B, 8:00 a.m.; L, 1:00 p.m.; D, 8:00 p.m.

"I'M NEVER DIETING AGAIN"

"When I say I've tried every diet out there, I mean it," said **Ann, the Wolf.** "Of course, I never lost weight. My pattern was that I'd stick with it during the day and then I'd start snacking on whatever food I was allowed to eat—low-carb, low-fat, the diet of the month—at night. I get it; my willpower was exhausted after a day of being 'good.' The simple idea of eating, in decreasing amounts, every four hours, whatever food I want, and then stopping after a late dinner, has been a revelation. I just don't get hungry at night if I have a late dinner. I was locked in a mind-set that I had to eat with my kids at 6:00 p.m. A good mother has a family dinner. But as soon as I decided to let that idea go, and eat my meal later while my husband had his dessert, my overeating just kind of stopped. I had my last bite, and that was it for the day. It's been a month. I have had a few nights of snacking, but nothing like before. And I have lost five pounds! Without really trying! I'm never dieting again. It seems like a miracle that I could lose weight just by eating later, but I'm not questioning it. I'm just doing it."

HAVE A DRINK

Failure: Getting inebriated, suffering miserably with a hangover, and doing damage to your master inner clock, your liver, and your GI tract.

Success: Indulging without getting drunk, lessening the severity of hangovers, sustaining minimal damage to your organs and master clock.

The Simple Science

I'm not a big drinker myself, in part because I know what alcohol does to your bio-time. Imagine putting a clock on the rocks (yes, I *can* pun), then taking a bottle of wine and smashing the clock. Excessive drinking shatters your inner clock to bits, and your body won't be able to set it right again for a long time, if at all. I'm sorry, but the research can't be denied. As I've previously explained, your body is full of biological clocks that keep your systems and organs on a tight schedule. Alcohol blurs many of these schedules. First, booze knocks out the SCN, the master clock located in your brain, and then it seeps down, knocking out other clocks that control, for instance:

- **The sleep/wake cycle.** Drinking in the evening suppresses melatonin release. In a Brown University study,[6] researchers monitored the sleep of twenty-nine healthy men and women in their early twenties for ten days. For three nights, half of the subjects drank a placebo beverage, and the other half got vodka one hour before bed. The vodka drinkers' salivary melatonin was reduced by up to 19 percent. Just one drink can cut your melatonin drastically.
- **Digestion.** Your second brain gets stupid when you drink, too. Since your master clock is affected, your guts don't know when they're supposed to do their job releasing proteins and enzymes. The unpleasant result is called "gut leakiness"[7] or "leaky gut syndrome." The bacterial lining of your gut fails to serve as a barrier, so bacteria, viruses, and toxins can leak into (and out of) the intestines, causing bloating, gas,

inflammation, headaches, skin conditions, food allergies, fatigue, and joint aches.

- **Liver function.** Your body's filter and metabolizer operates on its own clock, releasing proteins and molecules on a schedule. Alcohol disrupts that schedule, and in turn, the organ's mitochondria lose flexibility and start to break down, which causes liver disease.[8]

Just to be clear: You don't have to be an alcoholic to do long-term damage to your body and to dismantle circadian rhythms. **Chronic use, defined as two drinks a day, is enough to impair your master clock and, by extension, the 100 known minor clocks running on carefully calibrated timing throughout your body.** I recommend limiting alcohol consumption to four drinks per week *maximum* for men and women. You'll find higher numbers if you look for them. My recommendation is all about alcohol's impact on circadian rhythms, and a drink every other day is erring on the side of caution. Anyone at risk for GI tract, liver, and immune-related diseases and depression should abstain completely.

REALITY CHECK

Four drinks per week? How about four drinks per day? I know that many of you enjoy a glass of wine or two every night with dinner or indulge in a six-pack while watching football or at a party. People drink. They drink a lot. I'm not judging anyone for having a good time, or socializing with friends. My point is that drinking has consequences besides the ones your parents and gym teachers warned you about. It disrupts your circadian rhythm. The more you drink, the worse it is for your inner clocks. The closer you drink to your bedtime, the worse your quality of sleep will be, and the more impaired you'll be the next day by sleep deprivation, which compounds hangover symptoms. If this awareness makes you have one or three fewer drinks per week, great. If it inspires you to switch to seltzer at 9:00 p.m., great. Or not. If you're old enough to drink, you're old enough to make decisions and to be accountable for them. I'm just serving up the information. It's up to you whether you swallow it.

Attention, brunch Bloody Mary lovers and lunchtime martini drinkers: Consider **the tolerance rhythm.** At certain times of the day, the effects of alcohol are more pronounced. Back in 1956, when standards for experimenting on college students were a bit more lax, researchers at Stanford University Medical School[9] administered whiskey and water cocktails (20 percent alcohol) to their six subjects hourly (including a 3:00 a.m. dose). "The objective was, by repeated ingestion of small doses of alcohol, to achieve a slowly rising concentration in the body, the rise being slow enough so that an experiment could be continued for as long as 48 hours without the onset of severe degrees of intoxication," wrote the lead researcher, Roger H. L. Wilson. The researchers closely monitored the sipping subjects and measured their alcohol levels and rate of metabolism per hour via blood and saliva samples.

The subjects' ability to metabolize alcohol and clear it out of their system was highest in the evening and lowest in the morning.

Had the subjects been given a larger dose of whiskey, they would have been sloshed at 10:00 a.m. but mildly buzzed at 8:00 p.m. It is as accurate today as it was sixty years ago when Wilson et al. did their research. Human dehydrogenase is the enzyme that breaks down toxins including alcohol, and it has a circadian rhythm of its own. **Call it the bio-time happy hour.** Have your wine, beer, or cocktail when those enzymes are flowing in the early to mid-evening—enjoy the drink with friends, and have a conversation without slurring your words.

Of course, if you're setting out to get inebriated, have two mimosas at breakfast, and you will be blotto for the rest of the day.

NIGHTCAPS

If you are using alcohol as a sleep aid, you will wind up more exhausted for your efforts. Although it might work to make you fall asleep (or, if you drink enough, pass out), your sleep will be of inferior quality. An alcohol-addled brain gets stuck in Stage 3 and Stage 4 deep delta wave sleep and doesn't cross into REM sleep, which is when mental restoration and memory consolidation occur. Additionally, if you're inebriated when you fall asleep, you're more likely to sleepwalk, sleep-talk, sleep-shop, and sleep-eat. All you should do while sleeping... is sleep. At last call, have a seltzer.

And then there's the morning after. Alcohol disrupts bio-time. So, for starters, you'll wake up feeling social jet lag fatigue, fogginess, and irritability. **The hangover rhythm** is on the clock, too, in that the only cure for an aching head is the passage of time, rehydration, and doing whatever you can to get back on bio-time, like exposing yourself to direct sunlight, even if it's excruciating to do so.

Researchers at Kent State University[10] divided their hamster subjects into three groups: water drinkers, 10 percent alcohol solution drinkers, and 20 percent alcohol solution drinkers. After being exposed to dim light, the control group was bright and bushy and ready to run on their little wheels an hour earlier than their normal wake time. The 10 percent alcohol drinkers? They couldn't be roused until forty minutes after the control. The 20 percent alcohol drinkers? They took an hour longer still to wake up.

Many of us have witnessed the human equivalent in friends. The severely hungover shun light and, as if it were the middle of the night, can't get out of bed for many hours past their usual wake time. Heavy drinking in effect turns bio-time upside down. Hangover symptoms are very similar to those of social jet lag. Like an erratic sleep and wake schedule, hangovers are the result of messing up your circadian rhythm. The symptoms are nearly identical.

Sorry, Wolves. The chronotype most likely to indulge in alcohol is also the type to suffer the most extreme circadian disruption and frequent hangover symptoms. Researchers at the University of Barcelona[11] studied 517 students and found that evening types were most likely to partake of addictive substances, have alcohol consumption problems, and deal with hangover symptoms, including headache, hypersensitivity to sound and light, tiredness, anxiety, and irritability.

The Rhythm Recap

The tolerance rhythm: When alcohol is metabolized the quickest, so you can drink more with fewer effects.

The hangover rhythm: When excessive consumption causes social jet lag symptoms.

THE WORST TIME TO HAVE A DRINK

Brunch and last call. Early-morning drinking will get you drunker quicker—and sicker quicker. Late-night drinking will disrupt your inner clock and ruin the quality of your sleep.

THE BEST TIME TO HAVE A DRINK

Dolphin: Bio-time happy hours, **6:00 p.m. to 8:00 p.m.**

Lion: Bio-time happy hours, **5:30 p.m. to 7:30 p.m.**

Bear: Bio-time happy hours, **6:30 p.m. to 8:30 p.m.**

Wolf: Bio-time happy hours, **7:00 p.m. to 9:00 p.m.**

HAVE COFFEE (OR MY FAVORITE FORM OF CAFFEINE—CHOCOLATE!)

Failure: Drinking coffee first thing in the morning or too late in the evening, increasing caffeine tolerance, and bringing on insomnia.

Success: Timing coffee breaks with dips in cortisol levels to effectively boost energy.

The Simple Science

Caffeine is a drug—a legal stimulant. Coffee is a tasty delivery system and is available in nearly every home and on any city street corner all over the world. Somehow, it's become a cultural norm to associate coffee with waking up. There's a TV jingle that says so: "The Best Part of Wakin' Up is Folgers in Your Cup." People program their coffeemakers to start brewing when their alarm goes off so they can roll out of bed and right into a pot of Costa Rican brew.

I'm about to say something that will be a shock to many.

Drinking coffee first thing in the morning does not wake you up, make you alert, or give you an energy boost. All it does, according to science, is raise your tolerance for caffeine so that you need to drink more and more of it to feel any effects at all.

When you are about to wake up, your body releases stimulants to get your juices flowing and your heart pumping: a brew of hormones including insulin, adrenaline, and cortisol. Like most of our organs and glands, the adrenal gland (producer of adrenaline and cortisol) has a biological clock of its own. It carefully maintains **the cortisol rhythm,** a few cycles of releasing and suppressing the fight-or-flight hormone over the course of the day.

- **If you drink coffee when cortisol level is high,** the effects are nonexistent. Compared to cortisol, caffeine is weak tea. The only thing coffee does for you within two hours of waking is to increase your tolerance for caffeine.

- **If you drink coffee when cortisol level is low,** caffeine gently nudges your adrenals to give you a hit of adrenaline, and you will feel more awake and alert.

Scientific research and bio-time have provided a very clear schedule for coffee breaks to coincide with cortisol level dips. For the average Bears, those dips occur **between the hours of 9:30 a.m. and 11:30 a.m., and between 1:30 p.m. and 5:30 p.m.** (For your chronotype's optimal coffee window, see "The Best Time to Have Coffee" on page 208.)

Not to suggest that you should drink copious amounts of coffee during *every* cortisol level dip. If you do that, you'll develop a tolerance for caffeine, and you won't get the benefit of taking the drug. Researchers at the University of Oklahoma tested the cortisol response in a double-blind study[12] of nearly a hundred male and female subjects. For five days, the subjects abstained from coffee and took pills that were either a placebo or varying doses of caffeine three times a day. On day six, both groups were allowed to have coffee again, and their saliva was tested for cortisol levels. The control group, after having coffee again, showed a "robust" increase in cortisol levels. The caffeine pill group? Their cortisol response to coffee was reduced or eliminated.

What this means: The law of diminishing returns applies to drinking coffee three times a day. The more you drink, the less effective it is. If you're drinking coffee just because you love the taste, then switch to decaf to prevent developing a tolerance and fatiguing your adrenal glands.

When you drink more than 500 mg of caffeine per day (five cups of coffee, two energy drinks, ten sodas, or a combination of the three), you will feel nervous, restless, and cranky, and you'll have an upset stomach, muscle tremors, and heart palpitations. Chronically drinking too much caffeine can burn out your adrenal glands to the point where they can't produce enough cortisol on their own. The symptoms of adrenal fatigue are exhaustion, weight gain, memory loss, anxiety, low sex drive, chrono-misalignment, and associated risk factors and symptoms including depression, obesity, and heart disease.

And insomnia. Researchers from the Universidad de San Martín de Porres, in Lima, Peru, studied[13] 2,581 undergraduates and discovered a strong link between eveningness preference, daytime fatigue, and stimulant use. Wolfish students reported feeling tired during the day. They reported using coffee, cigarettes, and alcohol more than the morning and in-between types. Their stimulant use correlated directly with daytime fatigue. The more coffee they drank, the sleepier they were during the day.

That's probably because drinking coffee interferes with your sleep/wake cycle. **The melatonin rhythm** is largely controlled by the rise and fall of the sun, but it can be thrown out of sync by depressants, like alcohol, and stimulants, like coffee. To find out the extent of various substances' effect on melatonin secretion, a team of researchers conducted a forty-nine-day study[14] in which subjects were exposed to bright light or dim light or a double espresso before bed. Although bright and dim light suppressed melatonin, **coffee had the most significant effect, blocking melatonin for forty minutes, enough to ruin a night's sleep and cause chrono-misalignment.**

Patients tell me all the time that caffeine doesn't affect them, and that they can fall asleep after drinking coffee. I explain that they are probably so sleep-deprived that their brain throws them into sleep, but that with melatonin "off" due to caffeine, the sleep quality is inferior.

The plan is to have your last cup of coffee with sufficient buffer so you don't suppress melatonin. **The metabolism rhythm** is the length of time it takes to clear caffeine out of your system. I've previously discussed that it takes your body one hour to metabolize a glass of wine or a bottle of beer. How many hours does it take to metabolize the caffeine in one cup of coffee? One hour? Two? Four?

It can take up to forty-five minutes before you feel the effects of coffee, but it usually kicks in after twenty-five. Then, **it can take between six and eight hours for the stimulant effects of caffeine to be reduced by half.** You will feel your morning cup until afternoon. If you have an afternoon cup, you'll feel both until evening. And if you have another cup after dinner? You will be caffeinated way past your bedtime.

Researchers at Wayne State University in Detroit tested[15] the caffeine effect by giving subjects 400 mg of it right before bed, three hours prior, and six hours prior. Compared to the placebo control group, all three of the dose-time groups had significant disturbances to their sleep.

My advice? Don't drink any coffee until your cortisol level dips. Have your last cup before 2:00 p.m. (Dolphins and Wolves) or 3:00 p.m. (Lions and Bears).

The Rhythm Recap

The cortisol rhythm: When your adrenal gland releases and suppresses cortisol; this occurs in cycles throughout the day.

The melatonin rhythm: When excessive caffeine consumption is disrupted by caffeine intake.

The metabolism rhythm: When your body clears the caffeine out of your system.

THE WORST TIME TO HAVE COFFEE

Within two hours of waking. Within six hours of bedtime, especially if you have sleep problems, stress, or are a Dolphin.

THE BEST TIME TO HAVE COFFEE

The cortisol level dips or optimal coffee break times for each chronotype are:

Dolphin: 8:30 a.m. to 11:00 a.m.; 1:00 p.m. to 2:00 p.m. No caffeinated beverages after 2:00 p.m., including decaf coffee (yes, there is caffeine in decaf).

Lion: 8:00 a.m. to 10:00 a.m.; 2:00 p.m. to 4:00 p.m.

Bear: 9:30 a.m. to 11:30 a.m.; 1:30 p.m. to 3:30 p.m.

Wolf: 12:00 p.m. to 2:00 p.m. No caffeinated beverages after 2:00 p.m., including decaf.

REALITY CHECK

What? Take away your wake-up cup of joe? You might be thinking I'm some kind of evil doctor right now, and not one who's trying to help you. People push back on delaying coffee far more than any other change I recommend in this book—including alcohol restriction and early weekend wake time—and I'm not even suggesting you quit drinking it! I'm only saying, hold off for a couple of hours. Or, hold off for half an hour for a few days. Then an hour, then an hour and a half, and so on.

You drink coffee upon waking out of habit, because marketers and advertisers have convinced you to associate waking up with it. But any sleep expert or endocrinologist knows that caffeine does *not* make you less sleepy first thing in the morning. It only makes you jittery. Save your caffeine for when it's useful! If you must drink coffee before you open your eyes, have decaf. I *promise*, you will prefer how it makes you feel.

WHAT ABOUT CHOCOLATE?

I've worked with hotels to create sleep kits for their guests. It's a little bag with earplugs, an eye mask, some calming lavender spray, a night-light so they don't have to turn on the lamp to find the bathroom, and a curtain clip for blocking out ambient light. A sleep kit like this is a much better idea than what most hotels put on your pillow at nighttime: a piece of chocolate. One little piece isn't too bad. But eat a few pieces, and the caffeine in the candy could disrupt your night.

We've all switched to dark chocolate from milk chocolate because it has more antioxidants. Great for us. But dark chocolate also has more caffeine. A two-ounce chunk of 70 percent dark chocolate has 70 mg of caffeine, about the same as a cup of espresso. Chocolate contains theobromine, which is good for blood pressure because it widens your blood vessels. On the other hand, dilated vessels stimulate the heart and make you have to urinate—not ideal when you are trying to sleep.

Don't have your chocolate dessert after dinner. Instead, indulge in the morning, and you'll have the whole day to burn those calories and metabolize the caffeine. Also, the health benefits of chocolate are negated if you eat so much of it that you gain weight.

"COFFEE WITH LUNCH"

"If I had to pick the two changes that had the biggest impact on my happiness, I'd say morning sex and afternoon coffee," said **Ben, the Bear.** "Morning sex has been a surprise for my wife and me, because we thought we didn't have time before work and the kids' school. But we're like teenagers, being sneaky about it, the intensity. Sex has become a fun, exciting part of our mornings, instead of what it used to be, the lazy pre-bed routine. Also, drinking coffee at lunch instead of breakfast has been huge. I used to drink two cups at breakfast and still feel groggy, so I'd have another cup at work. (Sex is my new wake-up drug of choice!) Now I have one cup with lunch, and it perks me up for the entire afternoon. I'm drinking one-third the coffee and getting ten times the benefit of the caffeine. Anyone reading this book has to try this. If I crave the flavor and smell of coffee in the morning, I have decaf."

PIG OUT

Failure: Overeating and converting excess calories into fat, muddling your cognition the day after.

Success: Occasionally overeating and converting the calories into energy to fuel your body and mind.

The Simple Science

It happens to all of us, standing at the fridge in sweatpants with a fork and a guilty conscience. Why does the urge to consume junk food strike most often at night, and what is the impact of eating out of sync with your rhythms?

Your chronorhythm has a **willpower rhythm.** Research has explained why you can be "good" all day and then throw caution to the wind at night. In one famous study,[16] subjects were split into two groups—one whose members could eat cookies anytime they wanted, and another whose members were instructed to resist. Then both groups were given mind puzzles to solve. The first group, the cookie eaters who didn't have to exercise willpower, did better on the puzzles than the second group, those who had to resist the cookies. Practicing willpower drains mental energy. **You can only resist for so long before you exhaust your "being good" capacity.**

Compounding that is **the appetite/artificial light rhythm.** There's a reason humans reach for unhealthy food at night. In 2013, researchers from Oregon Health & Science University conducted a study[17] of a dozen non-obese adults, monitoring them in a dimly lit room in a lab for thirteen days. The researchers recorded the subjects' appetites and food cravings for the duration. Their meals were served at regular intervals, but the subjects chose what foods and how much to eat. At 8:00 a.m., they ate the smallest amounts. At 8:00 p.m., their appetites increased, and they consumed large quantities of starchy, salty, and sweet food.

Why were they pigging out at night? Leptin, the satiety hormone, the one that makes you say, "I couldn't eat another bite," is lowest in the afternoon (kudos to cultures that eat their biggest meal at lunch) and highest at night. You shouldn't have an appetite after dark, and yet people do. Why? Artificial light, the kind that comes from lamps and screens, is to blame. In the Oregon study, the subjects lived in dimly lit rooms. In an Ohio State University study,[18] nighttime exposure to bright or dim light caused a significant increase in body mass index in mice. The control group, exposed to a normal cycle of light and darkness, experienced no increase.

Late-night bingeing has an impact on **the memory rhythm,** too. Researchers from the University of California, Los Angeles, conducted a study[19] with mice to see if disrupting mealtimes (and hence circadian rhythms) affected memory. One group was fed at regular, normal times, reinforcing a healthy circadian rhythm. Another group was allowed to eat only during sleep hours. Both groups were given the same number of pellets and logged the same total hours of sleep, albeit on different schedules. Three weeks later, **researchers ran memory tests on the mice. The day-eaters far outperformed the night-eaters.** The second group couldn't remember objects or tones. Researchers put the mice in MRI machines and noticed physical changes in the learning and memory centers of the night-eaters' brains after only a few weeks of feeding off-schedule.

The Rhythm Recap

The willpower rhythm: When your ability to resist tempting food is diminished over time.

The appetite rhythm: When hunger and cravings are turned upside down by nighttime exposure to artificial light.

The memory rhythm: When eating out of sync with bio-time impairs memory and cognitive function.

THE WORST TIME TO PIG OUT

9:00 p.m. to 5:00 a.m. This applies to all chronotypes, especially Wolves, who are hungry at midnight. Eating a high-carb, high-calorie meal after dark causes weight gain and muddles memory. Eating early in the day lowers BMI and speeds up metabolism.

THE BEST TIME TO PIG OUT

Lion: 2:00 p.m. To combat an afternoon energy dip, pig out on protein.

Dolphin: 10:00 a.m. or brunch time.

Bear: 8:00 a.m. Make breakfast the biggest meal of the day, and the time to eat too much.

Wolf: 8:00 a.m. Make breakfast the biggest meal of the day, and the time to eat too much.

SNACK

Failure: Mindless eating at random times between meals, converting the extra calories into fat.

Success: Mindful snacking at specific times, converting the calories into energy and alertness.

The Simple Science

If you go too long between meals, your blood sugar will drop and your fat/sugar metabolism will slow down. If you want to lose weight and maintain energy and alertness, you do not want that to happen. Snacking can keep your digestive and metabolic clocks in sync and operating at full capacity.

Snacking strategically with that goal in mind requires that you keep a snack schedule. Nibbling a handful of this and a bag of that off-schedule sends your gut and liver clocks spinning wildly out of bio-time. I know food is all around, all day long. **I am not going to tell you what you can and can't eat, only when. Snack on a schedule. If it's not time to snack, don't.** Follow this one time-wise tip, and you will lose weight and you won't feel hungry.

So when should you snack? Lions should follow **the between breakfast and lunch rhythm.** Since they have breakfast at 6:30, and a socially acceptable lunchtime is a long, long way off for them, **Lions should have a mid-morning snack around 9:30 a.m., along with coffee, if desired.**

Everyone else should follow **the between lunch and dinner snack rhythm.** A study[20] of 123 postmenopausal, overweight women in Seattle demonstrated that it was easier to lose more weight and eat more healthfully on that schedule than on other schedules. The subjects enrolled in a weight loss program and kept track of their eating patterns over a one-year period. Those who snacked in mid-morning lost significantly less weight than those who snacked in mid-afternoon (7 percent compared to 11.5 percent of their body weight). The mid-morning snackers also tended to have more than one snack per day. The mid-afternoon snackers had more fruits and veggies and fewer calories overall. Researchers drew a connection between morning snacking and mindless eating; the subjects weren't actually hungry, but they ate anyway. A good way to break the mindless eating habit is to avoid snacking entirely until afternoon.

What about a bedtime snack? **The after-dinner snack rhythm** depends on chronotype:

- **Dolphins:** If hungry, Dolphins can eat a 100- to 200-calorie snack that's 50 percent carbs, 50 percent fat (cheese and crackers; cereal with milk; apple slices with peanut butter, a handful of almonds) an hour before bed to help lower cortisol levels and boost serotonin levels.

- **Lions:** Late-night snacking isn't an issue for you. You're not hungry and don't need dietary help to ease you into sleep.

- **Bears:** I advise against bedtime snacks for Bears. According to a 2011 Brazilian study,[21] eating closer to bedtime negatively affects the onset and quality of sleep; this applies especially to women. The fifty-two subjects—healthy, non-obese nonsmokers—kept detailed food diaries for a few days and were then monitored in a lab overnight as they slept. The study's male late-night high-fat snackers spent less time in REM sleep and ranked lower in sleep efficiency (sleep vs. the frustration of just lying there) compared to their non-snacking counterparts. The females who ate late-night snacks, high-fat or otherwise, rated low in every sleep assessment category. It took them longer to fall asleep, to reach REM, and to stay asleep than it took female non-snackers. The more they ate, the lower their sleep quality. Bears snack at bedtime out of habit, and they should make every effort not to do so. Instead of eating, they can try drinking Banana Tea (see my own recipe below).

- **Wolves:** Wolves would be happy to have a full-sized meal at midnight. In another Brazilian study[22] of a hundred subjects, a majority of whom were middle-aged, overweight women, researchers tested morningness vs. eveningness preferences, the tendency to eat late at night, and binge-eating habits. The evening group was more likely to binge at night. Wolves are susceptible to this bad habit because nighttime is when they come alive and expend energy, and when they get cravings for high-fat food. A night eating pattern, regardless of type, will result in excess weight and could lead to disease and depression. My recommendation for Wolves is to have a very small snack one hour before bedtime. Try 100 calories of half-fat, half-low-glycemic carbs (nuts, veggies with hummus) to calm you down before bed and satisfy the urge to eat. Or have Banana Tea (recipe below) to quiet your nerves and put something in your stomach.

BANANA TEA

Most of us know that bananas have magnesium. You might not know that the peel has three times more magnesium than the fruit itself. Since magnesium helps calm the nerves, Banana Tea is the perfect nighttime drink.

Ingredients:

1 ripe banana
2.5 cups boiling water

Instructions:

1. Thoroughly wash the banana to remove dirt, bacteria, and pesticides. It's best to use organic bananas.
2. Cut ¼ inch off the top and bottom ends of the banana.
3. Leave the peel on and cut banana in half, lengthwise.
4. Place the two banana halves into boiling water and boil for ten minutes.
5. Strain banana water into a cup.
6. If desired, add a drop of honey or cinnamon before drinking.

The Rhythm Recap

The between breakfast and lunch rhythm: When Lions should snack for energy and to stay on bio-time.

The between lunch and dinner rhythm: When Dolphins, Bears, and Wolves should snack for energy and to stay on bio-time.

The after-dinner snack rhythm: When you risk throwing your bio-time out of whack.

THE WORST TIME TO SNACK

The middle of the night. If you eat 25 percent of your total caloric intake late at night or overnight, you might have Nocturnal Eating Syndrome (NES), an eating disorder. A checklist of symptoms:

- Grazing on small meals hour after hour late at night.
- Waking up a few times overnight to snack and go back to bed.
- Rising from bed in the dark to eat a few nights per week.
- Knowing you've woken up to eat again and again, and feeling guilt or shame about doing so.

If this sounds like you, I suggest you reach out to a sleep expert in your area to explore your therapeutic options. Go to www.sleepcenters.org to find one.

THE BEST TIME TO SNACK

Dolphin: 3:00 p.m.

Lion: 9:00 a.m.

Bear: 4:00 p.m.

Wolf: 4:00 p.m.

SNACKING AFTER EXERCISE

If you exercise intensely for an hour or more, you *must* snack within half an hour of completing your workout. A snack of one-third protein, one-third complex carbs, and one-third simple carbs (yogurt with granola and berries, a shake with whey protein, oatmeal and fruit) is the ideal formula to rebuild stressed-out muscles and encourage afterburn, the metabolic process of burning stored fat for energy for up to twenty-four hours post-exercise.

If you exercise at medium or low intensity for less than an hour, you should not snack afterward, especially if your next meal is coming up in an hour or two. Your body does not need more calories to compensate for what you've burned and probably doesn't require additional carbs and protein to replenish spent muscles. Instead of eating, hydrate with water. Skip the energy drinks that are full of sugar, artificial sweeteners, and sometimes caffeine.

Work

ASK FOR A RAISE

Failure: Approaching the powers that be about a pay increase when they're distracted, in a foul mood, or too tired to deal.

Success: Approaching the powers that be about a pay increase when they're alert, in a good mood, and eager to hear your argument.

The Simple Science

Only you and your colleagues know the culture and climate of your particular workplace. Some offices or businesses are slow in the morning and busy in the afternoon, or vice versa. Your boss might be too busy to talk to you at certain hours or on certain days. You have to tune in to the unique rhythms of your place of business and become a student of your boss's moods and patterns. Figure out when he or she is generally receptive and available.

Obviously, don't bring up a fat raise when your company is in financial collapse (in that case, start interviewing for a new job ASAP; more on that subject in "Go on a Job Interview," page 237). Be aware of established performance review schedules and salary policies. At some workplaces, you could hang your company's logo on the moon and still not be eligible for a raise until the next calendar year.

If your business is financially sound, with flexibility about salary increases—and you've been working hard and doing well—by all means, start planning what you're going to say to the boss about why you deserve more money. Have your arguments prepared beforehand, of course. And walk in there at a right time—when the boss will be open to having the conversation from the get-go.

Some broad-strokes time-wise tips on **the DOW rhythm**. I don't mean the cycles of the Dow Jones Industrial Average. I'm referring to the day of the week.

- **Monday:** *Back off.* With the new week looming, it's likely the boss is sleep-deprived (due to Sunday Night Insomnia or social jet lag),

220

irritable, and stressed out. Unless you come in with ideas to make his or her life easier, do not go in there.

- **Tuesday:** *Steer clear.* Research suggests that Tuesday is the most hectic day in the workweek. With so much going on, your boss won't necessarily be interested in discussing non-urgent concerns. In fact, she might get a little annoyed that you'd interrupt the office productivity with personal matters.

- **Wednesday:** *Still not ideal.* Wednesday is the second most hectic day of the week. Your putting individual concerns above the needs of the team might come off wrong. Instead, be efficient and effective. Set the table for your just deserts.

- **Thursday:** *Don't jump the gun.* Although Thursday is the second *least* productive day of the week, it's also the day when people are the least positive and happy about being there.[1] In that context, asking for *anything* might not be met with an enthusiastic response.

- **Friday:** *Go for it.* The least productive day of the week, and the best day in terms of a positive outlook. People are happy about the upcoming weekend and not crazy-busy. Friday is your day to pounce.

Next, take into account your boss's **pleasantness rhythm**. In psychology, we use the term "positive affect" to mean good mood. When is the boss's positive affect the highest? For that matter, when is his or her grumpiness high point (or peak "negative affect")?

Back in 1995, psychologists from the Johannes Gutenberg Universität, in Mainz, Germany, set out to determine the day of the week and time of day that people were in the best mood. Forty-nine subjects rated their mood three times a day for an entire week on two measures of positive affect, including pleasantness (feeling balanced, happy, content, and at ease). This study, remarkable in its simplicity, corroborates the long-held belief that mood is amplified in *the middle* of the waking circadian cycle. **The German study found that people are grumpiest in the morning, get progressively happier in the afternoon, and are happiest in the evening.**

Extrapolating by chronotype:

- **Dolphin pleasantness peak: 4:00 p.m.**
- **Lion pleasantness peak: 2:00 p.m.**
- **Bear pleasantness peak: 6:00 p.m.**
- **Wolf pleasantness peak: 8:00 p.m.**

If you can determine your boss's chronotype, coordinate your request with his or her peak pleasantness and, therefore, peak receptivity to your request.

And when are you in the best frame of mind to do the asking? This is where **the activation rhythm** comes in. Another subcategory of positive affect is called activation (feeling attentive, active, interested, and inspired). Activation is exactly what you need to march into the boss's office and state your case.

For a study[2] by researchers at Harvard and Georgetown University, subjects were given cash incentives to guess people's weight in photographs. They were allowed to take the advice of anonymous influencers who rated their own confidence after their weight guesses. Regardless of the accuracy, subjects put their money on the advisors who rated their confidence at 100 percent.

For you to ask for a raise, you need to rate your confidence at 100 percent or to be at peak activation. The activation pattern is not identical to pleasantness. According to that German study, **activation is at medium intensity in the morning, at peak in the afternoon, and lowest in the evening**.

Extrapolating by chronotype:

- **Dolphin activation peak: 4:00 p.m.**
- **Lion activation peak: 12:00 p.m.**
- **Bear activation peak: 2:00 p.m.**
- **Wolf activation peak: 5:00 p.m.**

The Rhythm Recap

The DOW rhythm: When, in any given office environment, it's most likely that a private discussion about salary will be appreciated.

The pleasantness rhythm: When your boss is in the best mood to discuss your salary.

The activation rhythm: When you are full of confidence.

THE WORST TIME TO ASK FOR A RAISE

Monday morning. If you march into the boss's office first thing Monday morning and ask for more money, you'll be lucky to leave the room with your head.

THE BEST TIME TO ASK FOR A RAISE

Thursday or Friday afternoon. To fine-tune the timing by chronotype, I've created the chart below to find the nexus of your boss's pleasantness peak and your activation peak, favoring the boss's mood as the more important factor. Also, this doesn't take into account what's happening at work, the deadlines and crises that might pop up. And, since no one wants to be asked for anything with one foot out the door, ask at least one hour before the end of the workday.

Ask for a Raise Compatibility Chart

You ↓	Dolphin Boss	Lion Boss	Bear Boss	Wolf Boss
Dolphin	4:00 p.m.	3:00 p.m.	5:00 p.m.	5:00 p.m.
Lion	3:00 p.m.	1:30 p.m.	3:00 p.m.	3:30 p.m.
Bear	3:30 p.m.	2:00 p.m.	4:00 p.m.	4:00 p.m.
Wolf	4:30 p.m.	3:30 p.m.	4:30 p.m.	5:00 p.m.

COLD-CALL

Failure: Reaching out by phone to a stranger, acquaintance, or indirect contact to make a pitch...and getting swatted down—or you get them on the line and it doesn't go well.

Success: Reaching out to a stranger, acquaintance, or indirect contact to make a pitch...and getting him or her on the line or a call back.

The Simple Science

If you are just starting out, breaking into a new field, or looking for a new job, you will have to call people you don't know well and ask them for an interview or for advice. If you work in sales, personal finance, or real estate—any profession that requires you to generate business—cold calling is a way of life. *Everyone*, at some point in his or her career, has had to dial a stranger and make an elevator pitch.

Some chronotypes are better at cold calling than others. The three personality traits that are required for successful cold calling—which involves a lot of frustration, rejection, and disappointment—are resilience, optimism, and persistence.

First, **the resilience and optimism rhythm:** According to a Spanish study[3] from the University of Málaga, chronotype, resilience, and optimism are correlated. Not a huge surprise: Morning types (Lions) had the highest resilience and optimism scores; "neither" types (Bears) were in the middle; evening types (Wolves) had the lowest scores. The authors of the study noted, "These results suggest that evening-type subjects could display less capacity to face adversity and adapt positively, as well as less expectance of the occurrence of positive events compared to neither and morning-type individuals." In other words, Wolves aren't predisposed to cold calling.

But what about **the persistence rhythm?** You might not feel positive about what you're doing, but if you have grit, you can keep doing it anyway, right? Well, it's possible, but according to the research, Wolves don't score too well in persistence, either. In a German study[4] of chronotype and temperament, morning types scored highest in persistence and cooperation; "neither" types scored in the middle; evening types scored the lowest.

But that doesn't mean Wolves are doomed to fail at cold calling. They might not be personally suited to it, but they can use time-wise tricks to better the odds of success. The question, as always, is when to place those calls to increase the likelihood that the person on the other end will pick up the phone and be willing to give you five minutes.

You have to take into account the bio-time of the person you are trying to reach. **The weekly mood shift rhythm,** or the best day of the week to dial, is all about recovery from a weekend's social jet lag symptoms.

For a 2010 comprehensive survey by marketing research organization Lead Response Management, James Oldroyd, PhD, now an associate professor at Brigham Young University, studied three years' worth of data from half a dozen companies making over 100,000 sales call attempts. He was able to calculate which days of the week a caller is likely to reach an actual human and turn that connection into a possible sale.

- **Monday:** *Put the phone down.* Monday was a horrible day to connect. The research doesn't say why, but I believe the culprits are Sunday Night Insomnia and Monday social jet lag. People are irritable, tired, and grouchy and not in the mood to deal with a pitch yet. (Did you read "Ask for a Raise" on page 220? Similar principles are at work here.)

- **Tuesday:** *Don't even.* Tuesdays were the worst day to connect. Social jet lag is like crossing time zones. Generally speaking, it takes one day to recover for every hour crossed, or an hour of delayed sleep due to sleeping in. If a Bear, for example, slept until 10:00 a.m. on Sunday — three hours of lag time — he won't recover from social jet lag until Wednesday.

- **Wednesday:** *Dial away.* Wednesdays were the second best day to connect. No wonder, since most people have finally recovered, are back on bio-time, and are a lot less irritable and tired.

- **Thursday:** *Keep dialing.* You are 49 percent more likely to connect on Thursday than on Tuesday. It's the best day to get a human on the phone.
- **Friday:** *Maybe.* You might get someone on the phone, but this is the worst day to get what you were after (an interview or a sale). With one foot out the door for the weekend, people might take the call to kill time, but they are unlikely to commit to anything.

Data was also calculated to determine **the daily mood shift rhythm** for successful cold calling, or the best time of day to connect with a person:

- **7:00 a.m. to 8:00 a.m.:** Very early morning is the worst time to connect, and it's not hard to understand why. Only Lions are in work mode at that hour.
- **8:00 a.m. to 10:00 a.m.:** First thing in the morning is a great time to connect due to morning procrastination. Dolphins, Bears, and Wolves aren't yet at their peak focus and concentration, so they'll pick up the phone and waste time chatting for a minute.
- **10:00 a.m. to 2:00 p.m.:** The heart of the workday is *not* a good time to connect, especially if reaching out to Bears. They're at peak performance from 10:00 a.m. to 2:00 p.m. and are focused on their own work.
- **2:00 p.m. to 4:00 p.m.:** Bad time to call, due to the post-lunch cortisol level dip that puts everyone (except Lions) into an afternoon lull.
- **4:00 p.m. to 6:00 p.m.:** The hands-down best time to connect is in the late afternoon, when the post-lunch dip has subsided. Dolphins, Bears, and Wolves have an uptick in energy and mood and will be open to listening to your pitch.
- **6:00 p.m. to 7:00 p.m.:** Don't bother. Obviously, people are ready to leave the office, if they haven't left already.

The Rhythm Recap

The resilience and optimism rhythm: When to use positive affect to bounce back from rejection and believe the next call will be a winner.

The persistence rhythm: When to use determination to keep going, no matter what.

The weekly mood shift rhythm: When social jet lag makes cold-call recipients more or less likely to pick up the phone on certain days of the week.

The daily mood shift rhythm: When daily fluctuations in hormones and body temperature make cold-call recipients more or less likely to pick up the phone.

THE WORST TIME TO COLD-CALL

12:00 p.m. to 2:00 p.m. The biggest takeaway of this section: Do not pester people at lunch with a sales call. A social call? Sure.

THE BEST TIME TO COLD-CALL

Since cold calling requires determination and concentration—if someone answers, you have one minute to get your point across—it should be done during on-peak alertness hours that coincide with the availability of the person you're calling. This means that you should call from 8:00 a.m. to 10:00 a.m. and from 4:00 p.m. to 6:00 p.m.

Dolphin: 4:00 p.m. to 6:00 p.m., your optimal "in the zone" time.

Lion: 8:00 a.m. to 10:00 a.m., your optimal "in the zone" time.

Bear: 4:00 p.m. to 6:00 p.m., when you hit a second wind of alertness and are in a good mood and can handle rejection better.

Wolf: 4:00 p.m. to 6:00 p.m., when you are just getting warmed up.

WHEN TO CALL BACK

Say you've made a dozen calls and are waiting to hear back. You step away from your desk or put down the phone, and come back to find a voicemail with a message to call back. When should you return the call?

Immediately.

Oldroyd determined that if you can get the person on the phone within five minutes, you'll be twenty times more likely to actually talk to a human than if you waited thirty minutes. If you wait for longer than sixty minutes, the likelihood of connecting decreases tenfold. Maybe this is due to our ever-shrinking attention spans? If you don't pounce on a hot prospect ASAP, the lead will go cold after just one hour. My advice is, check your voicemail often, and don't silence your phone during work hours!

COMMUTE

Failure: Getting to work when you're prone to traffic accidents, wasting time on the road/train.

Success: Getting to work when driving is safer, making good use of the time on the road/train.

The Simple Science

You live where you live, and work where you work. Unless you move or change jobs, there's not a lot you can do about the miles between those two places, and the amount of time it takes to travel between them.

In an ideal world, all companies would have flexible hours and offer telecommuting instead of requiring employees to be on-site at a specific hour. Commuting brings us down. A Canadian study[5] of 3,409 daily drivers to work found that the longer their commute, the lower the subject's self-reported life satisfaction. The drivers complained of "time crunch" pressure, saying that they were rushing through life and constantly under stress. But those who worked in companies with flexible hours rated themselves higher in overall happiness and outlook. **The flexibility rhythm** for commuting is especially crucial to Wolves and Dolphins, who tend to rate life satisfaction as low. **If you think you can convince your boss to give you flexible hours, by all means, go for it. Otherwise,** until the day you land a job at Google, you'll have to figure out other ways to cope with the stress of getting to work before you're awake and alert.

It's important that Wolves and Dolphins develop coping mechanisms; if they don't they might put themselves and others at risk. **The vigilance rhythm** is when you drive most safely. In a recent Spanish study,[6] researchers looked at the differences in driving performance (reaction time, cautiousness, confidence, comfort behind the wheel) according to chronotype and time of day. The participants were assessed for chronotype and then took a simulated driving test at 8:00 a.m. and again at

8:00 p.m. When evening types drove in the morning, they made more errors than at night and were less vigilant. The morning types, naturally more cautious, were good, stable drivers, making fewer errors than the evening types overall and staying vigilant about safety both in the morning and in the evening tests.

- **Wolves and Dolphins:** If at all possible, take public transportation or have someone else drive in the mornings.
- **Lions and Bears:** If at all possible, take public transportation or have someone else drive in the evenings. Yes, Lions, you are vigilant behind the wheel at all times. But accidents do happen, most often during the evening commute. (See "Drowsy Driving" box below.)

Or, as an alternative, to improve overall life satisfaction, you can follow **the active commuting rhythm** by walking or biking to work. In a study[7] of 18,000 British commuters conducted by Norwich Medical School, University of East Anglia, England, researchers found that those who switched from passive (car) to active (walking, biking) travel greatly improved overall psychological well-being, general health, and concentration capacity when they got to work. Morning exercise helps clear sleep inertia; Dolphins, Bears, and Wolves will arrive at work in a clearer state of mind than if they'd been lulled on a subway or train. The study subjects claimed that their increased fitness and the happiness of active commuting knocked out the negatives of possibly greater commuting time. Time-wise tip for Wolves and Bears: If you walk or bike to work, you gain an hour in the evening by doing your exercise first thing in the morning.

The Rhythm Recap

The flexibility rhythm: When travel to work is on a flexible schedule to improve overall life satisfaction.

The vigilance rhythm: When to commute by car to make the fewest driving errors and be most cautious about safety.

The active commuting rhythm: When to walk or bike to work for better health and mental clarity and save an hour you would have spent exercising later on.

THE WORST TIME TO COMMUTE

Evening rush hour: 6:00 to 9:00 p.m. Most weekday traffic accidents occur during this block of time.

THE BEST TIME TO COMMUTE

Most of us don't have a choice about when we have to be at work. But if you did have a choice, the best time to commute would be when you were alert for driving and on an even keel for dealing with the stress of rush hour.

Dolphin: Morning commute, 9:30 a.m.; evening commute, 6:30 p.m.

Lion: Morning commute, 7:00 a.m.; evening commute, 3:00 p.m.

Bear: Morning commute, 8:30 a.m.; evening commute, 5:30 p.m.

Wolf: Morning commute, 11:00 a.m.; evening commute, 7:00 p.m.

DROWSY DRIVING

The three DDs of car accidents are drunk driving, distracted driving, and drowsy driving—getting behind the wheel when you are fatigued or foggy with sleep inertia. According to AAA Foundation of Traffic Safety statistics as of November 2014, drowsy driving is accountable for:

- **31 percent of all fatal accidents.**
- 13 percent of all accidents that send a person to the hospital.
- 7 percent of crashes that result in a car being towed from the scene.
- 6 percent of crashes that require medical treatment for injuries.

This is not to suggest that traffic accidents are most common during the morning commute. On the contrary, **the majority of fatal car accidents**

happen during the *evening* commute, per the National Highway Traffic Safety Administration, because commuters feel time-crunched and guilty about not spending as much time with their family and friends as they'd like, so they speed.

TIME-WISE TIPS TO PREVENT DROWSY DRIVING ACCIDENTS

- Spray peppermint aromatherapy in the car, eat peppermint candy, or chew mint gum. The smell is revitalizing.
- Get some exposure to bright lights.
- Do some light exercise at rest stops, such as push-ups against the car. It gets blood flowing and will wake you up. Stopping for exercise will block "road hypnosis," when you're zoned out behind the wheel.
- Listen to a comedy podcast. Laugher and active listening increase alertness.
- Tickle the roof of your mouth with your tongue. Just try it.
- Nap-a-latte: Drink six to eight ounces of lukewarm or iced coffee and then take a twenty-minute nap. The nap decreases sleep drive. When you wake up, the caffeine will have kicked in.
- Drive with a partner.
- Take a nap, if possible, before you get on the road.

EMAIL

Failure: Crafting poorly written emails and sending emails when they'll be ignored or overlooked.

Success: Crafting well-written emails and sending them when they'll be read.

The Simple Science

Like most of us, I receive scores of emails every day—some that I asked for or appreciate receiving, plus a lot of junk. I do my best to reply to important emails and to read all the newsletters I subscribe to. But I just don't have time during the weekday to give many important emails the attention they deserve. After work, after the kids and my wife have gone to bed, I often go into my study to tackle my in-box. By then, it's late and I want to go to bed, so I rush through my replies, and wind up trashing a lot of stuff I wish I had time to read. I suspect that most of you can relate to this pattern of procrastination and deletion. Checking your in-box is both a way to waste time and a constant reminder that you have a lot to get done.

The procrastination rhythm, or when you scroll through your in-box but put off going through it until later (which may or may not ever happen), differs from chronotype to chronotype. (So is pretty much everything in life, as you have realized by now.) Which type is most likely to have thousands of unread, undeleted emails in their queue, and which type has an empty in-box and well-organized saved email folders? Or, which type procrastinates and avoids making decisions, and which type doesn't?

- **Dolphins** have historically self-identified as procrastinators due to their neurotic perfectionism, which can delay the completion of a task.
- **Lions** self-identify as non-procrastinators because they tackle hard tasks in the morning.

- **Wolves** self-identify as procrastinators, mainly because they can't get much done in the mornings.
- **Bears** fall in the middle because they delay a bit in the morning but then do the heavy lifting in the afternoon.

In a DePaul University/Universidad Complutense de Madrid study[8] of the correlation between chronotype and two types of procrastination—indecision and avoidance—researchers found that, among 509 participants:

- **Lions** rated low in avoidance and higher in indecision.
- **Wolves** rated high in avoidance and lower in indecision.

Bears strike the good balance between efficiency and ruthlessness, replying and trashing in an orderly flow. Dolphins get a bit neurotic about the pileup in their in-box and are compelled to keep it well ordered by deleting and putting emails into folders.

Lions, however, might put off replying to and trashing emails because they aren't sure how they want to handle it. Wolves are fine with making snap decisions about deleting, but they'd rather not deal with actually doing it until later.

Bears and Dolphins win in-box maintenance.

The writing rhythm, or when you should craft emails, depends on whether you're sending a professional communication or a personal one. Professional emails should be written at optimal times, when your mental clarity is peaking. When I open a short email, I tend to reply immediately. If I open a long email, I ignore it and hope to deal with it later. At on-peak alertness times, you won't veer off-topic and will stay focused, concise, and to the point. I read one study[9] that demonstrated that the ability to perform any cognitive task requiring vigilance (being careful and thoughtful)—like writing a concise, to-the-point email—is keenest for each chronotype at its overall optimal times. Lions will be most precise writing emails in the morning. Wolves will write at their lean and mean best in the evening. Dolphins' optimal time is in the late afternoon, and Bears' cognitive peak is in the late morning to early afternoon.

Personal emails to friends and family are best written at off-peak alertness times, when you're more likely to ramble and do a time-consuming photo edit or comment at length on a link.

Your well-crafted emails are more likely to be opened and replied to if you hit send at strategic times. **The sending rhythm**, or when to send an email to push the odds of its being read and replied to faster than it otherwise might be, has been studied by marketing research firms for over a decade. The research firm Yesware analyzed **500,000 sales emails** in the first quarter of 2014. According to its findings:

- **Weekend emails are opened and replied to at a higher percentage than weekday mailings, due to decreased "in-box competition."**
- **Early mornings and late nights get the highest percentage of reads and replies.**

Those stats refer to emails sent from a business to a person—say, from Pottery Barn to you, announcing a sale on bedding. What about professional emails from one person to another person? In 2015, a team of researchers from the University of Southern California's Viterbi School of Engineering and Yahoo! Labs conducted the largest study[10] to date about behavioral patterns for sending and replying to personal messages in the age of email overload. Some of their findings:

- **If they are going to reply at all, 90 percent of people will do so within one day of receiving an email. Half will reply within one hour.**
- **Reply time is faster in the afternoon and evening and slower late at night and early in the morning.**

As you can see, the best open rates for business emails (early morning; late night) and the fastest reply rate times for personal emails (afternoon and evening) do *not* match up. This makes intuitive sense to me. I'm more likely to click through to browse sofas and chairs when I'm off the clock, and to send replies about personal/professional matters during the business day, when my recipients are not, themselves, off the clock and browsing sofas and chairs.

The Rhythm Recap

The procrastination rhythm: When certain chronotypes are likely to avoid clearing out their in-box or are unable to decide how to reply to emails.

The writing rhythm: When to write a concise email to a professional contact; when to send a freer email to a friend or family member.

The sending rhythm: When to send an email so that the recipient opens it and replies.

THE WORST TIME TO EMAIL

For professional emails: very late at night.
For personal emails: mid-morning and afternoon.

THE BEST TIME TO EMAIL

Dolphin: Professional, **4:00 p.m. to 6:00 p.m.** Personal, **9:00 a.m. to 12:00 p.m.**

Lion: Professional, **7:00 a.m. and 10:00 a.m. to 12:00 p.m.** Personal, **3:00 p.m. to 5:00 p.m.**

Bear: Professional, **10:00 a.m. to 2:00 p.m.** Personal, **4:00 p.m. to 6:00 p.m.**

Wolf: Professional, **4:00 p.m. to 7:00 p.m.** Personal, **10:00 a.m. to 12:00 p.m.**

GO ON A JOB INTERVIEW

Failure: Showing up for a job interview foggy, scattered, and irritable.

Success: Showing up for a job interview mentally sharp and in a good mood.

The Simple Science

First impressions are *not* everything. We all know colleagues who made a poor first impression but, over time, proved to be smart and capable. But they *have* a job. In a competitive job market, you won't get a second chance to prove yourself. The first job interview matters. If possible, schedule it on bio-time.

What does your future boss or human resources manager want to see in a new recruit? That's easy: a smart, capable, energetic, confident, and eager candidate. Interviewers are also looking for a fun, positive person who will fit into the office culture. Many of us spend more time with our colleagues than our families. Employers don't want to hire someone they wouldn't want to hang out with—they'd prefer to think, *I'd like to have a beer with him.*

To that end, what is each chronotype's **likability rhythm?** Lions' positive affect is highest mid-morning, so they'll be happiest at 12:00 p.m. Bears are at their friendly best at 6:00 p.m. Dolphins' positive affect kicks in around 4:00 p.m. And Wolves? Their mood peak comes later in the day, at around 6:00 p.m.

It bears mentioning (no pun intended) that Bears' and Lions' positive affect amplitude (meaning the intensity of their good mood) runs higher than Dolphins' and Wolves', according to a University of Pittsburgh study[11] of 408 nondepressed adults. So a Wolf at her happiest is not as happy as a Lion at his happiest. When scheduling that job interview, Lions and Bears have a wider range of likable times (mornings into afternoon). Wolves and Dolphins have a smaller window (late afternoon, early evening) and won't shine quite as brightly as Lions and Bears.

Not to worry! Likability will only get you so far in life or on the job hunt. In weighing the impact of all the rhythms in this chapter, I think it's the least important. Likability is the cherry on top. But it's not the reason you'll get hired. So what is?

Potential employers are interested in the trio of traits that measure your ability to get things done: (1) alertness, (2) orientation, and (3) attention. Alertness (responsiveness, readiness) is a gimme. Being fuzzy around the mental edges and slow on the uptake won't recommend you to any employer. Orientation means having an understanding of the situation—being able to quickly figure out how things work when you walk into an office. In an interview, you have to be tuned in to what your future boss is saying. If your mind wanders, you lose. Each of these three traits operates in a distinct neural pathway in your brain, but, taken as a whole, they can be summed up in the phrase "executive functioning"— what any executive expects to see in a potential employee.

Researchers at Pennsylvania State University studied[12] **the executive functioning rhythm** by dividing eighty subjects by chronotype and assessing their alertness/orientation/attention. The subjects were given tasks and tests every four hours from 8:00 a.m. to 8:00 p.m. on the same day (no doubt a very long day for them). Predictably, morning types performed better early in the day, and evening types did better later in the day.

- **Alertness** shifted depending on time of day. Lions' and Bears' scores increased at the end of the day, surprisingly, while the Wolves' scores were static. In subjects' self-assessment of alertness, however, all types reported an increase as the morning progressed, with Lions and Bears falling off later in the day and Wolves continuing to rise. What this tells me is that Lions and Bears *underestimate* their alertness in the afternoon, and Wolves *overestimate* theirs.

- **Orientation** didn't change over the course of the day for any chronotype. You either have observational skills and can summon them at will or you don't and can't.

- **Attention** was adequate for all types by the end of the morning, with predictable increasing and decreasing focus as the day wore on.

It's wise, when scheduling a job interview, to consider the interviewer herself. Be aware that she is following **the ranking rhythm.** A psychologically insightful study[13] by researchers at the Wharton School of Business and Harvard University evaluated time-of-day patterns in the approval rankings of 9,000 MBA candidates. They found that the interviewers practiced, subconsciously or consciously, "narrow bracketing," or the judging of candidates based partly on how they had ranked candidates seen previously in the day. For example, if an interviewer gave the first four candidates a high rating, they might give the fifth a lower ranking to distribute judgments of the entire group, even if all the candidates were equally qualified. What does this have to do with timing? If you are one of the last interviewees, you are likely to get the short end of the stick. With this in mind, **it's preferable to be interviewed early in the hiring process, and early in the day.**

The Rhythm Recap

The likability rhythm: When to wow a future boss with your good mood and positive vibes.

The executive functioning rhythm: When you are at your most alert and attentive.

The ranking rhythm: When an interviewer is most likely to put you at the top of his or her ranking system.

THE WORST TIME TO GO ON A JOB INTERVIEW

After a bad night's sleep. If you're sleep-deprived, your likability and executive functioning will be dialed down or completely switched off. Anxiety about how you'll do can cause insomnia the night before, which is all the more reason to keep a tight sleep schedule all week long—so that any given night's deficit won't be too damaging.

THE BEST TIME TO GO ON A JOB INTERVIEW

Schedule job interviews at the earliest hour that you believe is likely to find you bringing your positive affect and executive functioning to the table. This would preferably be in the first half of the day.

Dolphin: 11:00 a.m.

Lion: 9:00 a.m.

Bear: 10:00 a.m.

Wolf: 12:00 p.m.

LEARN SOMETHING NEW

Failure: Taking in new information when your brain is like a brick wall.

Success: Taking in new information when your brain is like a sponge.

The Simple Science

Learning, like sleeping, digesting, and socializing, has a circadian rhythm. The learning rhythm is a U-shaped curve with peaks and valleys, just like the temperature and cortisol rhythms. Psychologists[14] have found that learning anything new—language or developmental skills, for example—is done most efficiently with a period of acquisition followed by a period of rest, followed by more acquisition. It's not, as traditional teaching would have us believe, a linear process of going from lesson one to lesson two, taking it all in at a steady pace. Humans are excellent learners, but we can take in only so much at a time before cognitive performance goes south.

The basic learning rhythm, or this U-shaped curve of cognitive performance,[15] follows a predictable pattern. For Bears, the first peak is from 10:00 a.m. to 2:00 p.m., when their intellectual capabilities are at the highest they will be all day. It falls into a valley after lunch, from 2:00 to 4:00 p.m. The second peak is between 4:00 and 10:00 p.m. The deepest valley of the day occurs from 4:00 a.m. to 7:00 a.m. If you try to learn during these valleys, you'll struggle to take in new information and probably won't retain it. But if you ride the crests of your learning waves, you'll have a greater capacity to absorb information—as long as you give your brain time to process what you've learned in those valleys.

Depending on your chronotype, the time of day, and the complexity of the material, different regions of your brain will be activated to take in the new information. **The brain region rhythm** was discovered by a team of Belgian and Swiss researchers[16] when they gave extreme morning and evening subjects increasingly difficult working memory tests—like remembering numbers, or repeating words in a certain order—twice a day. While the tests were being administered, blood flow to the subjects' brains was monitored using MRI machines.

When the subjects took the easiest tests, regardless of the time of day, they all got the same scores. But **when they took the hardest tests, morning types' scores dropped during the evening testing session. The same drop happened to evening types' scores during the morning session.** No big shocker there. What researchers took particular notice of, however, was how blood flowed in the subjects' brains and which regions each chronotype used when challenged. When Wolves took the hardest tests in the evening, their thalamic regions (in the midbrain and responsible for alertness and processing information) lit up more than did the morning types'. When Lions did the hardest tasks in the morning, their middle frontal gyri (in the cerebral region and responsible for executive decision making and higher information) lit up more than the evening types'. **Lions' and Wolves' brains rely on different regions when they are challenged to learn at their peak mental capacity. Chronotype isn't only about when you wake up and feel tired. It's about how your brain works.** (This facet of chronobiology isn't necessarily relevant to *when* each type learns—but it sure is cool.)

The sleep-deprivation rhythm can't be ignored, even if you're tired of listening to me talk about it. Dolphins know all too well that attempting to learn something new when sleep-deprived is close to impossible. Numerous studies[17] have linked insomnia with difficulties in every learning manifestation, including deteriorated working memory, long-term memory, attention, decision making, verbal functioning, and vigilance (the ability to keep at a task until it's done). Perhaps the saddest consequence of sleep deprivation is that motivation falls off a cliff. You are just too tired to get psyched to learn something new. The only way to overcome the sleep-deprivation rhythm is to get at least six (Dolphins) or seven (everyone else) hours of sleep per night, night after night. And when you do—and you will, if you follow my recommendations in "Go to Bed" (page 184) and "Wake Up" (page 167)—you will have adequate Stage 3, Stage 4, and REM sleep to consolidate what you've learned and to retain the information for future use.

The Rhythm Recap

The basic learning rhythm: When learning is most effective — achieved by alternating periods of information acquisition during on-peak times and resting during off-peak times.

The brain region rhythm: When different parts of the brain light up depending on chronotype and degree of difficulty of what you're learning.

The sleep-deprivation rhythm: When you can't learn new information because you're just too tired.

THE WORST TIME TO LEARN SOMETHING NEW

The middle of the night. The lowest learning valley on the U-shaped curve is between 4:00 a.m. and 7:00 a.m. You should be consolidating information then, not learning new things. If you're a college student pulling an all-nighter, use the wee hours for reviewing the material as opposed to memorizing new stuff.

THE BEST TIME TO LEARN SOMETHING NEW

Dolphin: 3:00 p.m. to 9:00 p.m., awake enough and at the peak of your learning curve.

Lion: 8:00 a.m. to 12:00 p.m., wide-awake and at the peak of your learning curve

Bear: 10:00 a.m. to 2:00 p.m., wide-awake and at the peak of your learning curve.

Wolf: 5:00 p.m. to 12:00 a.m., awake enough and at the peak of your learning curve.

MAKE A DECISION

Failure: Being unable to choose because you can't think clearly, making a risky snap decision based on an emotion.

Success: Making a well-thought-out call that you won't regret later on.

The Simple Science

We make thousands of decisions every day. Ideally, we'd make careful, rational calls about the many choices in our lives, especially when they have to do with our performance at work. If you make a bad decision that negatively affects your job and livelihood, it affects the rest of your life, too. If you use bio-time to choose wisely at work, you will benefit for the rest of your life.

Decisions are made rationally or emotionally, based on how the choice is framed. The "framing effect," a well-known concept in psychology, means that you're likely to make a choice depending on how it's presented to you. If a choice is framed as a potential gain, you're likely to avoid risk. If a choice is framed as a potential loss, you're likely to take greater risks. In emotional terms, if you feel confident and secure, you're cautious. But if you feel afraid and insecure, you're reckless. **The framing effect rhythm** is how you react (emotionally or dispassionately) depending on the time of day you have to make a decision.

In an Appalachian State University study,[18] subjects were asked to consider a classic example of framing called "the Asian disease question": A deadly Asian disease might kill six hundred people. You have two choices for saving some of them. Choice one guarantees that two hundred will live. Choice two offers no guarantees, with a slight chance everyone might live, and a higher chance that four hundred people will die. The framing effect at work here is in the language. Even though in both outcomes, two hundred people are likely to survive, choice one has a positive frame (risk free, two hundred will live!), and choice two has a negative frame (high risk, four hundred might die!).

Researchers asked this question to subjects at multiple times over the course of a day and found that there was a pattern to their responses. When the subjects were on-peak, they put aside the emotional reactivity and chose the logical, risk-free first option, that two hundred people would live. But when they were off-peak, they chose the emotional high-risk option, that it's possible everyone might live but that four hundred would probably die. In the movie of this scenario, the scientist character would push for the certainty of option one, but the renegade ex-soldier would push for the slight chance to save the world and go for option two.

I think we can all agree that it makes sense to use logical cognitive powers to choose rather than making snap emotional decisions. Next time you are faced with a quandary, consider waiting for on-peak times to make the call.

	Dolphin	Lion	Bear	Wolf
"On" Times to Make a Decision	10:00 a.m.–2:00 p.m.; 4:00 p.m.–10:00 p.m.	6:00 a.m.–11:00 a.m.; 2:00 p.m.–9:00 p.m.	8:00 a.m.–1:00 p.m.; 3:00 p.m.–11:00 p.m.	12:00 p.m.–2:00 p.m.; 5:00 p.m.–1:00 a.m.
"Off" Times to Make a Decision	9:00 p.m.–6:00 a.m.; 2:00 p.m.–4:00 p.m.	10:00 p.m.–6:00 a.m.; 11:00 a.m.–2:00 p.m.	1:00 p.m. to 3:00 p.m.; 12:00 a.m.–8:00 a.m.	1:00 a.m.–12:00 p.m.; 2:00 p.m.–5:00 p.m.

One warning about "on" and "off" decision times: Recall that sleep deprivation and sleep inertia are key factors. **The sleep-deprivation rhythm** formula: If you are fatigued or foggy, consider yourself "off." Your cognitive and performance abilities will be impaired. Making an important decision when sleep-deprived is as irresponsible as doing so when you're drunk. Wait until your head clears. How long that takes depends on the chronotype and the individual. For some (usually Lions), sleep inertia might last ten minutes. For others (Dolphins, Bears, and Wolves), it might last for up to four hours[19] after waking.

The personality rhythm comes into play when making decisions, too. Dolphins and Lions tend to be more cautious. Wolves and Bears tend to be more impulsive. Certain chronotypes are also likely to be indecisive and/or are likely to avoid pulling the trigger on a decision.

Any guesses?

You'll be surprised.

The procrastination rhythm is a particular problem for . . . wait for it . . . *everyone*, but in different ways.

You thought I was going to say "Wolves."

In a Spanish study[20] of 509 adults, researchers wanted to find out how different chronotypes handle "decisional procrastination" and "avoidant procrastination." They first tested to determine their subjects' chronotype and then assessed their particular procrastination patterns. The findings:

Morning types don't avoid making decisions. Using the example of shopping for a new outfit, Lions don't hesitate to go to the store. But once they get there, they have a hard time deciding what to buy.

Evening types, on the other hand, will avoid going to the store as long as they can, but once they get there, they don't hesitate to decide what to buy.

I can hear readers laughing with recognition.

It's scary how consistent the science is, isn't it?

The Rhythm Recap

The framing effect rhythm: When to make decisions based on logic vs. emotion.

The sleep-deprivation rhythm: When fatigue and sleep inertia affect decision making.

The personality rhythm: When certain chronotypes' cautious or impulsive nature leads them to make wise or snap decisions.

The procrastination rhythm: When certain chronotypes are indecisive or avoid having to make a decision.

THE WORST TIME TO MAKE A DECISION

First thing in the morning and the middle of the night. When you wake up, your cognitive powers and alertness are not up to speed. Wait an hour or more until your head clears to make a call. Your brain at rest should not be asked to do the mental processes required for rational deliberation. If the phone rings at 3:00 a.m. and you're asked to make a decision, put it off until morning. Unless you're a doctor or the president, the emergency can wait.

THE BEST TIME TO MAKE A DECISION

Dolphin: 4:00 p.m. to 11:00 p.m.

Lion: 6:00 a.m. to 11:00 a.m.

Bear: 3:00 p.m. to 11:00 p.m.

Wolf: 5:00 p.m. to 12:00 a.m.

MEMORIZE

Failure: Forgetting new information as it comes in, and not being able to recall information from stored memory banks.

Success: Being mentally prepared to take in new memories, lock them in, and recall them easily.

The Simple Science

Remembering the dialog from *Caddyshack* and recalling the names of the people you were just introduced to are both necessary tools for success in business and in your personal life. Whether it's long-term memory or short-term "working memory," you can sharpen them using time-wise tricks.

But first, a quick explanation of how memory works. Using song lyrics as an example, here's how the three-step memory-making process works:

1. **Acquisition.** You hear "Hello" by Adele on the radio.
2. **Consolidation.** The words "Hello from the other side / I must have called a thousand times" solidify in your mind.
3. **Recall.** When you hear the song "Hello" twenty years from now, you'll be able to sing every line.

The acquisition rhythm depends on being well rested. You could listen to "Hello" ten times when you're exhausted, and your brain wouldn't be able to absorb the lyrics. Perhaps, in college, you pulled an all-nighter to memorize facts before a test, and then, when you sat down to take it, you couldn't remember a damn thing. Or you could but it was lost after the test. According to a study[21] by researchers at Harvard Medical School, a single night of diminished sleep interferes with memory acquisition.

The consolidation rhythm, when you absorb new knowledge, preserve it, and protect it from disruption or degradation over time, happens during sleep. Sleep not only secures memory for future retrieval but also appears to free up the learning centers of the brain to take in new batches of information the following day. It's like moving pennies off the counter into the piggy bank so the counter is clear for tomorrow's take.

You don't need a full night's sleep to consolidate memory. A nap will do. If you heard "Hello" only once and then took a nap, you would be able to sing the song—not as well as Adele, obviously, but most of the lyrics would come to you. New York University scientists studied[22] the neural activity of mice and found that **a period of sleep immediately after learning a new skill encouraged the growth of synapses in the brain that were specifically related to what you just learned.** Mice that slept after initial learning to perform a task did twice as well on the new task as non-napping mice. Although the process is still mysterious, scientists believe that memory consolidation occurs during deep delta sleep, or Stage 3 and Stage 4 sleep, and, less importantly, during REM sleep. Dolphins might need to relearn tasks and incoming information several times due to sleep-deprivation-related consolidation shortfalls.

Regarding **the recall rhythm,** or when you can pull the name of a movie actor out of your brain or sing every lyric of a song you haven't heard in years, research[23] indicates that retrieval of short-term and long-term memory is impaired by sleep deprivation. Among the well-rested, retrieval of long-term memories peaks in the afternoon. Brazilian researchers[24] divided subjects into morning, evening, and neutral chronotypes (Lions, Wolves, and Bears, respectively) and trained their memory by teaching them ten words. A week later (considered "long-term"), the subjects were quizzed on the word lists with decoy words thrown in to confuse them. **Regardless of when the three groups were trained and later tested, all of them had better recall in the afternoon.** Perhaps the explanation is that, by afternoon, sleep inertia is completely gone. It's easier to pull a fact out of your memory bank when it's not clouded in fog.

The Rhythm Recap

The acquisition rhythm: When your brain is well rested and able to take in new information.

The consolidation rhythm: When your brain is in Stage 3, Stage 4, and REM sleep, locks in memory, and clears the decks to take in more information the next day.

The recall rhythm: When your brain is warmed up and best able to remember information.

THE WORST TIME TO MEMORIZE

After a bad night's sleep and first thing in the morning.

THE BEST TIME TO MEMORIZE

For all chronotypes, a good night's sleep is essential for acquiring, consolidating, and recalling memories. Memory acquisition happens all day long. As for the other two hurdles of memory:

Dolphin: Consolidation, **4:30 a.m. to 6:30 a.m.** Recall, **3:30 p.m.**

Lion: Consolidation, **3:30 a.m. to 5:30 a.m.** Recall, **2:00 p.m.**

Bear: Consolidation, **4:00 a.m. to 7:00 a.m.** Recall, **5:00 p.m.**

Wolf: Consolidation, **5:00 a.m. to 7:00 a.m.** Recall, **6:00 p.m.**

PRESENT YOUR IDEAS

Failure: Public speaking when your energy is low and you can't hold your audience's attention.

Success: Public speaking when your energy and focus are high, fully engaging the audience.

The Simple Science

Some chronotypes thrive when they stand in front of a room of people or at the head of a conference room table to present their ideas (talking to you, Lions). Others would rather close their door and get their individual tasks completed (Dolphins), relocate every workplace meeting to the local bar for happy hour (Wolves), or are content to be in the audience or part of a group presentation (Bears).

Whether you like being in front of a room, at some point in your career you will have the spotlight turned on and you will have to present your ideas (and yourself) at a meeting or conference. If you are in the position to schedule the presentation, you can use a few practical timing and time-wise tricks to make it a big success.

First and foremost, **the attendance rhythm:** When to hold the meeting or presentation to ensure a decent-size audience? You can't be dynamic and successful if you're addressing an empty room. The British website WhenIsGood.net, a portal for scheduling meetings, analyzed[25] two million responses to over half a million suggested events to determine most popular availability times. Attention, Lions: In an office populated by Bears, Wolves, and Dolphins, your Monday at 9:00 a.m. presentation is not going to be well attended, even if you bring doughnuts.

* Respondents were *most* flexible about fitting in a meeting between 2:30 p.m. and 3:00 p.m. The second best time: 10:00 a.m. to 11:00 a.m.

- Respondents were the *least* flexible at 9:15 a.m. The second worst time: 12:00 p.m. to 1:00 p.m. (lunch).
- Tuesday afternoons got the highest flexibility scores, and Monday mornings got the lowest number of flexible times.

The energy rhythm—when you are energetic, focused, and alert—doesn't arrive for most chronotypes until mid-morning, at the earliest. Lions are ready to roll at 8:00 a.m., but the average Bear won't hit his stride until 10:00 a.m., when sleep inertia has faded and cortisol and adrenaline levels have reached their morning high point. Wolves and Dolphins won't be capable of energetically and clearly presenting their ideas (or reacting intelligently to others' ideas) until mid-afternoon.

Of course, hold a presentation at your peak energy and alertness. But also remember that you can only be as good as your audience's ability to engage. **The engagement rhythm,** or when you are most likely to have the audience in the palm of your hand, is hard to pinpoint. Within any given group of people, there will be a range of chronotypes. At any given time of day, some might be fully engaged, while others are drifting, daydreaming, or napping with their eyes (and possibly their mouths) open. Instead of worrying about the best time of day to present your ideas, concern yourself with the length of time it takes to do it.

According to the website MeetingKing, an audience's attention span is extremely limited. After half an hour, one in seven will zone out. After forty-five minutes, one in three people's eyes will glaze over. Communications researchers at Texas Christian University conducted a study[26] about students' "anxiety in listening." The subjects were asked to listen to information with the knowledge that they'd be tested on it later. After a short time, the subjects hit their "cognitive backlog," when they couldn't absorb one more statistic or grasp one more concept. How short? Twenty minutes. TED Talks are always under twenty minutes for this very reason. It's also why the chapters in this book are so short! I want you to take in the info, not glaze over it. Even if the subject matter and the presenter are fascinating, humans can absorb for only so long. In a too-long presentation (or chapter), the audience stops being engaged, and your brilliant ideas won't be appreciated at all.

The Rhythm Recap

The attendance rhythm: The time of day when the greatest number of people will be flexible about coming to a meeting or presentation.

The energy rhythm: When you are at peak energy and alertness.

The engagement rhythm: When your audience is capable of actively listening to your presentation and appreciating what you have to say.

THE WORST TIME TO PRESENT YOUR IDEAS

9:00 a.m., 2:00 p.m., and 6:00 p.m. In the early morning, people are dealing with sleep inertia and are not at their energetic peak. In the post-lunch body temperature dip, energy is also at a low point. At the end of the workday, people's focus is tapped out and their attention span is down (Wolves are the exception). Even if you're in the zone, your audience will be zoned out.

THE BEST TIME TO PRESENT YOUR IDEAS

Make your presentation short and sweet — twenty minutes or less. If you have the power to set the presentation time, choose a time during an on-peak energy and alertness window:

Dolphin: 4:00 p.m. to 4:20 p.m.

Lion: 10:00 a.m. to 10:20 a.m.

Bear: 1:40 p.m. to 2:00 p.m.

Wolf: 5:00 p.m. to 5:20 p.m.

"GOING LAST HURT ME"

"I'm in marketing, and my entire job is presenting ideas to clients, colleagues, and my bosses," said **Robert, the Lion.** "In a group presentation, my attitude has always been, 'I'll go last.' Saving the best for last, right? I know

I'm better than anyone else, and I'll blow them away in comparison. This 'attention rhythm' stuff about people tuning out after a long meeting and how they subconsciously discounted the last person really rang a bell for me. Looking back, I can see how going last hurt me. I was tired by then, too. I had to put all my energy into what I was saying, and, as a result, I wasn't paying close attention to [the audience's] reactions." Lions do tend to be introverted. They get in their own heads and locked on their own goals. He continued, "I also decided to cut my presentations in half. I usually over-prepare and over-research, and I wanted to show everything I've got. But that probably hurt me, too. So now I go first and keep it short, and it's been going really well. I have a new perspective on it: My presentations aren't about me. They're about how the audience engages with me."

Creativity

BRAINSTORM

Failure: Racking your brain for ideas, coming up empty, and then feeling anxious and insecure about it.

Success: Coming up with great ideas with minimal effort when your mind is relaxed.

The Simple Science

Whether you're an artist in need of an idea, a marketer in a meeting, or a parent with dinner to prepare, you are in need of a brainstorm. Lightning needs to strike in the gray matter, bringing you a brilliant idea out of nowhere.

To get in sync with **the connectivity rhythm,** you must first understand when and where ideas begin brewing. Creativity starts in the prefrontal cortex, right before you wake up and in the first hour of the day. A neurological storm is under way, with pathways and connections lighting up across hemispheres. In a 2013 study,[1] researchers examined MRI scans of the resting brain in the morning and evening. **In the morning, they found bilateral links in the medial temporal regions, meaning your mind is making all kinds of connections that trigger fresh avenues of thought. In the evening, MRI scans show frontal and parietal brain correlations, meaning the brain is busy retrieving memories and not creating new ideas.**

The other big opportunity to brainstorm is when you're in sync with **the distraction rhythm**. When you're fatigued, cognitively dull, and incapable of concise thought, ideas slip in. Harvard researcher Shelley Carson, author of *Your Creative Brain*, has been studying this paradoxical phenomenon for years. One would think that the ability to focus brings on ideas. But actually, as she told the *Boston Globe*, "Distraction may provide the break you need to disengage from a fixation on the ineffective solution." **Have you ever tried and tried to come up with a solution for hours and been frustrated, only to give up, take a break, walk the dog, chop vegetables, or take a shower and, in five minutes, come up with the perfect idea?**

These creative highs come during off-peak alertness lulls,[2] or what I call "moments of groggy greatness," when you're likely to have random conceptual notions that become the seed of a brilliant idea.

The novelty rhythm should be heeded, too. Dopamine, the "fun" hormone, floods your brain when you are enjoying new experiences that are creatively inspiring. **When you feel good or are having a good time in a new situation, relationship, or environment, dopamine pathways in the limbic system are flowing.** In this heightened state, you will free-associate into having fresh and original ideas. Tech companies seem to understand this concept. Google's campus in Mountain View, California, for example, has sports and gaming areas for workers to unwind and play to inspire ideas.

The REM rhythm is another way to spur on creativity. Recall that REM sleep makes up about 25 percent of your sleep. However, it is not sprinkled evenly throughout the night. Most of it occurs in the final third (in fact, the final hour) of your sleep cycle. **If you are not getting enough sleep, you are likely cutting off REM—and creativity!** During the last two hours of sleep and for the first half hour post-wake-up, your brain is bubbling with ideas. Paul McCartney famously composed the ballad "Yesterday" in a dream. He woke up and immediately wrote it down. He couldn't believe that he could write such a perfect melody in his sleep. But it makes sense. While he was consolidating memory and his prefrontal cortex was humming, his brain was both actively connecting and putting the notes in place. The old adage "sleep on it" really does work well for both brainstorming and creativity. Allowing yourself to get the right amount of sleep and waking up without an alarm can ensure that you are at your "groggy greatness" with some regularity.

The Rhythm Recap

The connectivity rhythm: When the two halves of your brain are talking to each other, inspiring loads of ideas.

The distraction rhythm: When you're a little bit tired and easily distracted, ideas can pop into your head unexpectedly.

The novelty rhythm: When you're having fun and seeking new people, places, and experiences, you'll get fresh, original ideas.

The REM rhythm: When you're dreaming and consolidating, your brain is bubbling.

THE WORST TIME TO BRAINSTORM

11:00 a.m. to 3:00 p.m. It's a sad irony that you're at your least creative at the heart of the workday, when brainstorming is demanded of you.

THE BEST TIME TO BRAINSTORM

Dolphin: 5:00 a.m. to 8:00 a.m., 2:00 p.m. to 4:00 p.m.

Lion: 4:00 a.m. to 6:00 a.m., 8:00 p.m. to 10:00 p.m.

Bear: 6:00 a.m. to 8:00 a.m., 9:00 p.m. to 11:00 p.m.

Wolf: 7:00 a.m. to 9:00 a.m., 10:00 p.m. to 1:00 a.m.

PLAY MUSIC

Failure: Freaking out with performance anxiety and then flubbing notes while playing your instrument.

Success: Feeling calm and cool before a gig and nailing the notes with consistency.

The Simple Science

Musicians amaze me. They perform the miracle of manipulating piano keys, guitar strings, and/or their vocal cords to produce notes that, in precise combination, communicate human emotions. If you've been moved to tears by a virtuoso, you know that playing an instrument is an art as well as a skill. Any skill that requires dexterous muscular control, from playing a sport to playing the tuba (lips are muscles, too), is affected by the player's circadian rhythms.

The virtuoso rhythm: The circadian rhythms of a musician's sensorimotor precision (the ability to hear, see, and move at the same time in careful coordination) were investigated[3] by researchers at the Institute of Music Physiology and Musicians' Medicine, in Hanover, Germany. They wanted to find out how time of play affected pianists' skills. Were Lion pianists able to be precise in the evening, when most concerts are played? How well did Wolves play at matinees, when their fine motor skills might not be sufficiently warmed up yet? To sort it out, the researchers first tested twenty-one professional pianists to determine their chronotype (all other factors, such as practice time and sleep duration, were the same). Then the pianists played two-octave C major scales at 8:00 a.m. and again at 8:00 p.m. in a lab on two different days. The scales for each player at each time were evaluated on accuracy (playing the right notes), evenness (consistency of intervals between notes), stability (how consistently the performer played at a given session), and velocity (how fast the performer hit the keys, and how loud the music was).

Although both morning and evening types played accurately and evenly regardless of the session time, stability was a different matter. The Wolves' timing was more stable during the evening session. The Lions' timing was stable at both sessions. The researchers took particular note of the fact that these were expert musicians who trained constantly. Yet **for the evening types, circadian fluctuations did put a dent in morning performance. This suggests that sensorimotor fluctuations aren't always overcome with practice, practice, practice.**

As for keystroke velocity, or how fast and loud the music was, both chronotypes hit harder in the evening session. This result is consistent with other fine motor skills (grip strength, for instance), which are significantly better later in the day for all types. **Whether you play piano or tennis, you'll have more power after dark, even if you're a Lion.**

A Dolphin musician client of mine told me that she prefers nighttime gigs to matinees, which is only natural for her type. She and I have done some work together to help her control her performance anxiety. Even though she's played professionally for decades, she still gets nervous before a show. Anxiety is a common trait among Dolphins, but performance anxiety is not exclusive to bad sleepers. **The butterflies rhythm** strikes all chronotypes and can be warded off with well-timed meditation or relaxation.

Psychologists at the University of Sydney, Australia, studied[4] the benefits of a slow breathing relaxation technique on forty-six professional musicians. Some of the players were assigned to do controlled breathing for five minutes starting thirty minutes before a concert, and others didn't make any changes to their usual routines. Compared to the control group, the controlled breathing group — in particular, those most severely affected by performance anxiety — self-reported significant reductions in pre-show butterflies.

At www.thepowerofwhen.com, you'll find audio instruction about progressive muscle relaxation techniques. Or you can try the 4-6-7 Breathing Method to quiet nerves before a gig, a date, a meeting, or anytime, as needed.

1. Breathe in for a count of four.
2. Hold your breath for a count of six.
3. Exhale for a count of seven.
4. Repeat for two minutes.

The Rhythm Recap

The virtuoso rhythm: When your chronotype has the greatest sensorimotor precision in terms of speed, accuracy, and intensity.

The butterflies rhythm: When you get nervous before a show and need to calm down.

THE WORST TIME TO PLAY MUSIC

There is no bad time to play music. Practice, practice, practice still applies. You might play better at different times throughout the day and night. But, in most cases, the more you play, the better you'll be. One caveat: Don't plug in the guitar or bang the drums in the middle of the night, or your neighbors will be very annoyed.

THE BEST TIME TO PLAY MUSIC

Dolphin: Showtime is at **8:00 p.m.** Pre-show relaxation breathing at **7:30 p.m.**

Lion: 2:00 p.m. and 8:00 p.m. Matinee or evening curtain, you're equally proficient. Pre-show relaxation breathing at **1:30 p.m. and 7:30 p.m.**

Bear: 2:00 p.m. and 8:00 p.m. to 11:00 p.m. if you're playing late at a club. Matinee or evening curtain, you're equally proficient. Pre-show relaxation breathing at **1:30 p.m. and 7:30 p.m.**

Wolf: Showtime is at **8:00 p.m. to midnight if you're playing late at a club.** Pre-show relaxation breathing at **7:30 p.m.**

PUT IT ALL TOGETHER

Failure: Being unable to connect seemingly unrelated things and see the big picture.

Success: Being able to connect seemingly unrelated things and see the big picture.

The Simple Science

Detectives on TV seem to be able to put it all together effortlessly. While questioning a suspect or flipping through a little notebook, they get a funny look in their eye and suddenly remember a tiny detail or piece of information they learned days before that reveals the murderer's identity. Benedict Cumberbatch in *Sherlock* puts his fingertips to his temples to switch on his computerlike brain and access every bit of information packed in there. He connects the dots, no matter how random they might be, and creates a road map in his mind that leads right to the murderer's door. It's *Elementary* (the ability, and that other Sherlock show).

In real life, connecting seemingly unrelated bits and pieces that bring about a flash of insight — if that's not creativity, I don't know what is — usually doesn't happen in the interrogation room, the lab, or the library at 221B Baker Street, London. For mere mortals, insight strikes when (and where) you would least expect it: first thing in the morning when you're still in bed.

The insight rhythm is as simple as bio-time gets: During the day, when you're awake, you take in thousands of sights, sounds, smells, words, ideas, and so forth. At night, when you're asleep, the hippocampus region of your brain cements the day's input in your memory banks and mixes it all up with old memories of experiences and sensations from the past. The new and old input bounce around and light up various connections that are revealed to you in the morning when you wake up, and get an aha! pop of insight, the very piece of the puzzle you needed to plot your novel, crack a code, or solve the murder.

One study[5] by researchers at McGill University and Harvard Medical School trained subjects on a complicated sequencing math problem that was devised to force them to make seemingly unrelated connections in their minds. The subjects were split into three groups to be retested after a delay of twenty minutes, twelve hours, or twenty-four hours. Furthermore, the twelve-hour delay group was divided into those trained in the morning and retested in the evening, and those trained in the evening and retested in the morning after a good night's sleep.

The group that showed the greatest improvement on their scores: the twelve-hour delay group that trained in the evening, slept, and was retested in the morning. A twelve-hour delay and a good night's sleep gave their brains time to incubate ideas about how to solve the problem. When they woke up and were tested, their morning brains provided the flash of illumination. While they weren't actively working on the problem, their resting minds did the hard work of connecting the dots for them.

My insomnia patients say things like, "I lie awake at night trying to figure things out." I reply, "Sleep on it! Your brain will sort it all out for you. You'll have your answers in the morning."

Don't believe that "sleeping on it" helps you solve puzzles and problems? You can see just how profound the insight rhythm is yourself by taking the Remote Associates Test (RAT) before bed and then retesting yourself in the morning.

RAT was created in 1962 by University of Michigan professor Sarnoff Mednick, PhD, and holds up as an excellent assessment of creativity and mental agility. How it works: Each RAT question[6] shows you three seemingly unrelated words and asks you to provide a fourth word that links them together:

Easy

Cottage / Swiss / cake _____ [7]

Cream / skate / water _____ [8]

Loser / throat / spot _____ [9]

263

Medium

Sage/paint/hair _____ [10]
Boot/summer/ground _____ [11]
Fly/clip/wall _____ [12]

Hard

Animal/back/rat _____ [13]
Foul/ground/mate _____ [14]
Tail/water/flood _____ [15]

Sarnoff Mednick teamed up with *Take a Nap!* author Sara Mednick (his daughter; science runs in the family!), to prove that REM sleep gives creativity a boost. For their study,[16] the Mednicks gave all of their subjects the RAT at 9:00 a.m. and a word analogy test at 9:40 a.m. At 1:00 p.m., the subjects took naps that did or did not include REM sleep or just spent quiet time without sleep to incubate insight. At 5:00 p.m., the subjects took another RAT of questions they'd seen before or hadn't been exposed to, followed by one more memory test at 5:40 p.m.

The REM rhythm—taking a nap that includes REM sleep—yielded a 40 percent improvement between the subjects' morning and evening RAT scores. The incubators' and non-REM nappers' morning vs. evening scores were the same or worse. Only REM naps (on average, ninety minutes long) enhanced the brain's remote associative networks, helping subjects link disparate ideas in inventive ways.

The Rhythm Recap

The insight rhythm: When your brain connects seemingly unrelated ideas into a sudden aha! moment.

The REM rhythm: When taking a ninety-minute nap that includes REM sleep helps the brain transition from incubation to illumination.

THE WORST TIME TO PUT IT ALL TOGETHER

10:00 a.m. to 2:00 p.m. for Lions and Bears; 4:00 p.m. to 11:00 p.m. for Dolphins and Wolves. Peak alertness hours are for accumulating input that will become future memories (or clues, if you will). Don't bother trying to put the pieces together then. Instead, soak up information and sensory experiences.

THE BEST TIME TO PUT IT ALL TOGETHER

Dolphin: 4:00 to 7:00 a.m., during REM sleep and for half an hour after waking. No naps for Dolphins!

Lion: 3:00 to 6:00 a.m., during REM sleep and for half an hour after waking. **3:30 p.m.,** after waking up from a ninety-minute nap.

Bear: 4:30 to 7:30 a.m., during REM sleep and for half an hour after waking. **4:00 p.m.,** after waking up from a nap.

Wolf: 5:00 to 8:00 a.m., during REM sleep and for half an hour after waking.

WRITE A NOVEL

Failure: Trying to fill the blank page but coming up with nothing.

Success: Filling the blank page during your creative peak and editing during your analytical peak.

The Simple Science

I am under no illusion that any part of the writing process is "simple." It's never easy to sit down and stare at a blank page or blank computer screen. The blinking cursor is *always* intimidating.

As scary as writing may be, good bio-timing can give you an advantage. **You can type your way to new plot twists and inventive dialogue when your brain is primed to create, and edit the rough pages when your brain is primed to be analytical and strategic.**

The creating rhythm is almost the opposite of the other cognitive functions. The others—memorizing, decision making, paying attention, and planning—are best done when you're most alert. But the function of "creating"—what you have to do to bring to life the world in your head—is done best when you're *least* alert. In a landmark study[17] conducted by researchers at Albion University, 428 college students were measured for their circadian preference and split into morning, evening, and "neither" groups, and were asked to solve three problems each in two categories—"insight" or "analytical"—during two different testing sessions, one in the morning and one in the afternoon. The students had four minutes to solve each set of problems.

The "insight" questions used in the study:

- **The prisoner problem:** A prisoner is attempting to escape from a tower. He finds in his cell a rope that is half long enough to permit him to reach the ground safely. He divides the rope in half, ties the two parts together, and escapes. How could he have done this?[18]

- **The fake coin problem:** A dealer in antique coins gets an offer to buy a beautiful bronze coin. The coin has an emperor's head on one side and the date 544 BC stamped on the other. The dealer examines the coin, but instead of buying it, he calls the police. Why?[19]
- **The water lilies problem:** Water lilies double in area every twenty-four hours. At the beginning of the summer there is one water lily on a lake. It takes sixty days for the lake to become completely covered with water lilies. On what day is the lake half covered?[20]

The "analytic" problems used in the study:

- **The age problem:** Bob's father is three times as old as Bob. They were both born in October. Four years ago, Bob's father was four times older than Bob. How old are Bob and his father?[21]
- **The bachelor problem:** Five men, Andy, Bill, Carl, Dave, and Eric, go out together to eat five evening meals (fish, pizza, steak, tacos, and Thai) on Monday through Friday. Each bachelor serves as the host at a restaurant of his choice on a different night. Eric has to miss Friday's meal. Carl hosts on Wednesday. The guys eat at a Thai restaurant on Friday. Bill, who hates fish, is the first host. Dave chooses a steak house for the night before one of the guys hosts everyone at a pizza parlor. Which guy hosted the group each night, and what food did he choose?[22]
- **The flower problem:** Four women, Anna, Emily, Isabel, and Yvonne, receive a flower from their partners, Tom, Ron, Ken, and Charlie. Anna's partner, Charlie, does not give her a rose. Tom gives a daffodil to his partner (not Emily). Yvonne receives a lily, but not from Ron. Which flower (carnation, daffodil, lily, or rose) is given to each woman, and who is her partner? [23]

As you can see, the "insight" questions are about being creative; the "analytical" questions are a mathematical grind-it-out process. The students got more answers correct in the analytical category during their optimal time based on chronotype. During non-optimal times, they solved more insight questions. This is why I recommend **writing during off-peak times**, and **editing during on-peak times.**

If you are doing other kinds of writing — technical, business, the less "creative" kinds — write and edit during on-peak times. If the job is to present the facts in a clear, concise, coherent way, use your focus and concentration powers to get it done without the distraction of a wandering mind.

You can force a non-optimal state of mind. How? **The intoxication rhythm.** Researchers from the University of Chicago did a study[24] called Uncorking the Muse, about the effect of alcohol on the creative process. Some subjects stayed sober, and others were given a vodka and cranberry juice cocktail. They drank to the blood alcohol content of .075. For a 160-pound man, it would take three and a half cocktails to get there. (By United States standards, legal intoxication is .08.) All of the subjects were given Remote Associates Test questions. The intoxicated subjects solved more of them in less time than the sober control group.

So when are you most susceptible to alcohol, if you intend to uncork the muse yourself? See "Have a Drink," page 200. (For the record and for obvious reasons, I don't recommend this, and definitely not within three hours of bedtime.)

The routine rhythm: Is that a bit redundant? I don't think so. Establishing a writing routine primes your creative mind to spark up at the same time each day. Instead of citing a study in this section, I'll let some famous writers speak to the benefit of sticking with a regular writing schedule. You could do a lot worse than taking advice from these award winners and best sellers:

When I'm in writing mode for a novel, **I get up at four a.m. and work for five to six hours.** In the afternoon, I run for ten kilometers or swim for fifteen hundred meters (or do both), then I read a bit and listen to some music. I go to bed at nine p.m. I keep to this routine every day without variation. **The repetition itself becomes the important thing; it's a form of mesmerism. I mesmerize myself to reach a deeper state of mind.**

—Haruki Murakami

When I am working on a book or a story **I write every morning as soon after first light as possible.** There is no one to disturb you and it is cool or cold and you come to your work and warm as you write. You read what you have written and, as you always stop when you know what is going to happen next, you go on from there. **You write until you come to a place where you still have your juice and know what will happen next and you stop and try to live through until the next day when you hit it again.** You have started at six in the morning, say, and may go on until noon or be through before that.

—Ernest Hemingway

I write in the morning and then go home about midday and take a shower, because writing, as you know, is very hard work, so you have to do a double ablution. Then I go out and shop—I'm a serious cook—and pretend to be normal. I prepare dinner for myself and if I have houseguests, I do the candles and the pretty music and all that. **Then after all the dishes are moved away I read what I wrote that morning.** And more often than not if I've done nine pages I may be able to save two and a half or three.

—Maya Angelou

I tend to wake up very early. Too early. Four o'clock is standard. My morning begins with trying not to get up before the sun rises. But when I do, it's because my head is too full of words, and I just need to get to my desk and start dumping them into a file. I always wake with sentences pouring into my head. So getting to my desk every day feels like a long emergency.

— Barbara Kingsolver

I need an hour alone before dinner, with a drink, to go over what I've done that day. I can't do it late in the afternoon because I'm too close to it. Also, the drink helps. It removes me from the pages. So I spend this hour taking things out and putting other things in. **Then I start the next day by redoing all of what I did the day before,** following these evening notes.

— Joan Didion

Like your bedroom, your writing room should be private, a place where you go to dream. **Your schedule—in at about the same time every day, out when your thousand words are on paper or disk—exists in order to habituate yourself,** to make yourself ready to dream just as you make yourself ready to sleep by going to bed at roughly the same time each night and following the same ritual as you go.

— Stephen King

CHRONOTYPES OF FAMOUS WRITERS[25]

Dolphin

Alexandre Dumas

Franz Kafka

Charles Dickens

Marcel Proust

William Shakespeare

Lion

Maya Angelou

W. H. Auden

Benjamin Franklin

Victor Hugo

John Milton

Toni Morrison

Haruki Murakami	Kurk Vonnegut
Flannery O'Connor	Edith Wharton
Sylvia Plath	

Bear

Jane Austen	Thomas Mann
Stephen King	Susan Sontag
George Orwell	

Wolf

Ray Bradbury	Gertrude Stein
F. Scott Fitzgerald	William Styron
Charles Darwin	Hunter S. Thompson
James Joyce	J. R. R. Tolkien
Kingsley Amis	Mark Twain
Vladimir Nabokov	Virginia Woolf

The Rhythm Recap

The creating rhythm: When to invent a world and characters in fiction during non-optimal times; when to edit your rough pages on optimal time.

The intoxication rhythm: When to force non-optimal creative flow by having a few drinks.

The routine rhythm: When to establish a writing routine to get to work and quit at the same times every day.

THE WORST TIME TO WRITE A NOVEL

2:00 p.m. to 3:30 p.m.; 12:00 a.m. to 7:30 a.m. The only times you *shouldn't* be writing or editing your novel? When you should be napping or sleeping, and consolidating memories and making remote associations that will help you to create the next day.

THE BEST TIME TO WRITE A NOVEL

Dolphin: Write, **8:00 a.m. to 10:00 a.m.** Edit, **4:00 p.m. to 6:00 p.m.**

Lion: Write, **8:00 p.m. to 10:00 p.m.** Edit, **6:00 a.m. to 8:00 a.m.**

Bear: Write, **6:00 p.m. to 11:00 p.m.** Edit, **10:00 a.m. to 2:00 p.m.**

Wolf: Write, **8:00 a.m. to 11:00 a.m.** Edit, **6:00 p.m. to 10:00 p.m.**

MORNING PAPERS

Julia Cameron, author of *The Artist's Way*, a 1992 self-help book about creativity that has sold millions of copies, urges people to write "Morning Papers," three pages of longhand stream-of-consciousness thoughts. As she explains in a video[26] on her website, Morning Papers can be "anything at all, like 'I forgot to buy kitty litter' or 'I need to wash the curtains.' They seem to have nothing to do with creativity, but what they do is clear your mind. It's as though you've taken a little dust-buster and you go poking it into all the corners of your consciousness and come up with what you put on the page."

Cameron advises her readers to let the thoughts flow uninterrupted, even if they go into some fuzzy or dark places. "A lot of times people think they should be artful. I say no. They should be whining, petty, and grumpy," she says. "You write down just what is crossing your consciousness. It's as though you're writing down in meditation terms what we call 'cloud thoughts' that move through your consciousness and becoming acquainted with the dark corners of your psyche. You are meeting your shadow and taking it out for a cup of coffee. What I find is that when you put the negativity on the page, it isn't eddying through your consciousness. Morning papers are a clearing exercise and an exercise that makes you have much more consciousness as you pass through your day."

The pages are *not* for your novel or for anyone's eyes but yours. They aren't poetry or journal entries. Your stream of consciousness is, as she said, a clearing process that actually buys you extra time during the day. She says it takes her thirty or forty minutes to accomplish this. Doing so makes you more efficient and motivated to action afterward, which is a net time-wise gain. I'm a supporter of any process that clears away negativity and gives you more time to succeed.

When to fit Morning Papers into your chronorhythm:

Dolphin: 6:30 to 7:00 a.m.
Lion: 6:00 to 6:30 a.m.
Bear: 7:00 to 7:30 a.m.
Wolf: 7:00 to 7:30 a.m.

Money

BUY

Failure: Wild overspending on stuff you don't really need, owing to emotional reactions to external and internal cues.

Success: Spending appropriately when you are able to make rational purchasing decisions.

The Simple Science

The obvious difference between shopping and buying is that only one of them involves the exchange of cash for goods and services. Theoretically, you could shop all day long and not spend a penny. There is no charge for browsing. Shopping can be pure entertainment, something to do to kill time, or it can be a pre-purchase exploration of what's out there.

The best time to shop is when you leave your wallet at home.

Trust me on this. Every cue in a retail environment — in-store and online — is devised to get you to take out your credit card *right now*. Outlets spend millions on research to figure out how to appeal to consumers' emotional instability and get them to throw rationality out the window.

Stores count on your not being able to resist. You are up against enticing "Super Sale!" and "Buy 2, Get 1 Free!" temptations that have you tossing random items into your cart without thinking. Bears and Wolves are more susceptible to **the impulse purchase rhythm** than cautious Lions and Dolphins. But in the right circumstances, anyone is vulnerable to making an on-the-spot purchase that will later be regretted.

This is especially true of nonmenopausal women during the luteal phase of their cycle (that is, between ovulation and menstruation). It's not sexism, it's science. According to a survey[1] of 443 women from eighteen to fifty, those in the luteal phase had less control over their spending and impulse purchasing than those in the follicular and ovulation phases. Plus, they didn't feel bad later on about having made those purchases, meaning that at the time, they had no spending control based on guilt and regret. Impulsive spending + permissiveness = debt. No judgment, just simple truths.

Women and men alike: When you are in a state of arousal—not sexual; excited, full of joy—you are likely to make an impulse purchase. For this reason, store environments are visually exciting (bright lights, colors, smells, music). You become "wide-eyed" in a figurative *and* literal sense: When you are aroused, your pupils dilate. In a state of heightened arousal, you become distracted, forget what you intended to buy in the first place, and are tempted by stuff that wasn't on your list.[2] Studies[3] have found that the wider your eyes, the more exciting the retail environment, the looser your wallet.

To avoid the impulse shopping rhythm, shop when you are least likely to become aroused by an exciting in-store environment. As I've explained before, there are three types of arousal: energy, tension, and hedonic tone (happiness), all of which play a role in impulsivity. According to a Polish study[4] of chronotypes and arousal:

- Evening types tend to be more anxious and less happy over the course of the whole day compared to morning types.
- Morning types tend to have more energy than evening types until 5:00 p.m.

In terms of the three-dimensional model of arousal . . .

Wolves, Dolphins, and later-rising Bears are less susceptible to in-store wide-eyed purchases during the daytime.

Lions and earlier-rising Bears will have greater wallet control in the morning but are susceptible in the evening.

Many people buy things not because they're excited but for the opposite reason: They are feeling down, and spending money makes them feel happier (if temporarily). **The retail therapy rhythm** *can* lift dark clouds from over your head, per a study[5] by researchers at the University of Michigan. Why? Life can make you feel out of control, but when you are the one making purchasing decisions, you feel more grounded. Wolves and Bears who need the occasional pick-me-up would do well to give themselves a limited budget, and to stop buying when they reach it.

The Type A rhythm is about the nexus of purchasing and personality. Ambitious types (Lions) are more likely to be materialistic and to view acquisition as a way to reach their personal goal of success. A study[6] of 193 Australians drew a direct link between materialism and the Type A personality traits of competitiveness and aggression. Objects—a new car, a big house, a diamond bracelet—are like the spoils of war to them. As a psychologist, I feel obliged to give a gentle warning about acquisition as a way to mark your status in comparison to others: You are not your possessions. Eventually, buying stuff to validate your accomplishments will feel hollow. That realization might not happen until you're seventy, though, so enjoy the toys in the meantime.

Dolphins are generally Type A personalities, too, but on the neurotic and anxious end of the spectrum. That can put a strain on their wallets. According to an Israeli study,[7] when materialistic people are under fire (the study's 139 subjects lived in a town that had sustained rocket attacks), they reacted by going shopping to relieve stress. Ultimately, buying things didn't make them feel happier or more secure, but they did it anyway.

One last thing to watch out for to keep your wallet intact: **the hunger rhythm.** As we all know, you should *never* go grocery shopping when you're hungry, or you'll wind up clearing the shelves. Researchers[8] from the University of Minnesota proved that being hungry causes shoppers to spend more on nonfood items, too. In lab experiments, hungry subjects were faster to grab stuff greedily, even inconsequential objects like binder clips. In field research, researchers polled eighty-one shoppers on their hunger level as they exited a department store, and then checked their receipts. The hungry shoppers purchased a greater number of nonfood objects and spent nearly twice as much money as those who weren't hungry, regardless of their mood or how long they'd been shopping.

The Rhythm Recap

The impulse purchase rhythm: When you're incited to buy stuff during highly aroused times (and when female shoppers are about to get their period).

The retail therapy rhythm: When you buy more to feel less sad.

The Type A rhythm: When you acquire objects as a way to measure your personal achievement; when you buy things at times of extreme stress or anxiety.

The hunger rhythm: When you spend more on an empty stomach.

THE WORST TIME TO BUY

Defining "worst" as when you're most susceptible to overspending:

- **Female shoppers:** A week before your period, two hours before a meal.
- **Male shoppers:** Two hours before a meal.

THE BEST TIME TO BUY

Defining "best" as when you're least susceptible to overspending, purchase when you're (1) not hungry, (2) in an energy lull, and (3) at an arousal low point. That is, buy after lunch.

Dolphin: 1:00 p.m.

Lion: 12:00 p.m.

Bear: 2:00 p.m.

Wolf: 3:00 p.m.

"AMAZON MUST THINK I DIED"

Dolphin Stephanie told me she used to spend part of every evening after dinner browsing websites. "I thought shopping was relaxing, but since I clued into the concept of shopping when stressed, I realized that my browsing happened only at night when I got antsy. It was a knee-jerk reaction. I'd feel nervous"—with an evening cortisol level surge—"and I'd go online. Now I use a web-blocker app that locks me out of shopping sites during those hours, and do other things that actually work to calm me down, like reading and taking a bath. I'm sure Amazon must think I died or something! I'll say this for the power of when: My Visa bill is a lot smaller."

GET RICH

Failure: Attempting to accumulate wealth using strategies that conflict with your chronotype's personality traits.

Success: Taking advantage of your chronotype's personality traits to grow rich.

The Simple Science

This is not a book about "how" or "what." For tips about investing, saving, and insider trading, you'll have to go elsewhere. Nor can I supply you with specific time-of-day or day-of-week tricks for building a portfolio or hedging (whatever that means). I can tell you that specific personality traits associated with each chronotype can be leveraged to your advantage in your pursuit of financial security.

For example, **the morning ritual rhythm** is a common practice of successful Lions. They get up early and make the most of the few hours of quiet time before the world wakes. A list of titans who rise before 6:00 a.m. includes the current or former CEOs of Apple, AOL, Bloomberg, Cisco, Condé Nast, Chrysler, Disney, General Electric, General Motors, Oxygen, PepsiCo, Procter & Gamble, Starbucks, Unilever, Virgin, and Xerox, to mention several. Why do the mega-wealthy put mind over mattress and get out of bed at the crack of dawn? They're driven to achieve. The urge to accomplish is in their DNA.

Lions tend to be more ambitious, competitive, and successful in the business realm than other chronotypes, and can often point to their morning time as the secret of their success. Business journalist Laura Vanderkam wrote an entire ebook on the subject, called *What the Most Successful People Do Before Breakfast*, and described the morning rituals of dozens of rich and powerful people. So what do they do when the rest of us are dreaming? According to Vanderkam, they exercise, work on a passion project, meditate, make to-do lists, spend time with family, strategize their big-picture plans, catch up on emails and information, and network with, we can assume, other Lions, to plot world domination.

THE BEST TIME TO DO EVERYTHING

"Early to bed, early to rise, makes a man healthy, wealthy, and wise" is a saying often attributed to Lion Benjamin Franklin, who was also a huge proponent of to-do lists and began his (early) morning by asking, "What good shall I do this day?" His 5:00 a.m. to 8:00 a.m. morning ritual was to "rise, wash, and address *Powerful Goodness*; contrive day's business and take the resolution of the day; prosecute the present study; and breakfast." Or, in other, more modern words, "get up, take a shower, meditate, plan what you're going to accomplish and vow to get it done, do a bit of catching up on what's going on, and then EAT!" Sounds like a plan.

Dolphins and Bears are not alert in the mornings, and forcing themselves through a Lion's morning ritual would be a waste of sleep time that they need to prevent deprivation. (Being sleep-deprived is NOT anyone's route to wealth.) Instead, those who need a few hours to get up to speed should pay attention to **the big idea rhythm**—or use prime creativity time to harvest brilliant ideas that will put dancing dollar signs in their eyes.

Before I get into specific "whens" of big ideas, a very brief explanation of where they come from. The three idea-generating brain networks are:

* **The Control Network.** This involves parts of the prefrontal cortex and the posterior parietal lobe. It lights up when you are focused and concentrating on challenging problem solving.
* **The Imagination Network,** or the default-mode network, involves parts of the prefrontal cortex, the medial regions of the temporal lobe, and other areas. It's the "wandering mind," the daydreaming network, when your thoughts meander into the past or project into the future.
* **The Flexibility Network,** in the dorsal anterior cingulate cortices and anterior insular regions, keeps your mind aware of what's happening inside and outside your body, and helps prioritize the most important thoughts you have at any given moment. It might be, "I'm hungry," or it might be to follow a thin thought thread from the Imagination Network to the Control Network to dream up and concentrate on a big idea.

Put all three of these networks together, and you'll be able to generate, recognize, and hone in on a big idea.

But when? *When* will the insight come? The truth is, it can happen at any time. For a study[9] of the neuroscience of insight, researchers at Northwestern University used EEG and MRI machines to pinpoint neural activity on different timetables during subjects' aha! moments. They found that sudden insight is exactly that: sudden. It can be totally unrelated to preceding thoughts, on its own timetable, unpredictable, and impossible to track.

That said, a wealth of other research[10] has determined that the wandering, distracted, non-optimal mind is primed for sudden insight. When your mind is "at rest," it is not idle. It's actively processing memories and ideas. Creative, golden haze thought does not occur while you're reading the paper, watching TV, checking email, or surfing the Internet. But it might occur when you're in the shower, chopping vegetables, meditating, staring into space, or lying in bed between snooze one and two. At Google, a company that appreciates the mechanism of creativity, creative thought might happen during what they call "20% time," the portion of some engineers' workday devoted to daydreaming to come up with new ideas.

Bears and Dolphins have many opportunities for productive mental drift throughout the day when they are not on task or concentrating. I recommend that Bears and Dolphins take a page from the Lion playbook, and make a list of specific time slots when they can daydream for twenty minutes or so without a thought agenda. It might be right after waking, or during the afternoon lull, or while taking a hot bath at night.

Wolves are naturally creative. If anything, they have too many bright ideas and not enough organization and follow-through. I've always thought that Wolves and Lions should be business partners. The Wolf can be the idea person, and the Lion is the operational person.

Wolves have another secret weapon for wealth, one that Lions don't have. Any guesses?

What *are* the habits of seriously wealthy people? Of course, networking, hard work, and diligence are important. But so is taking risks. Predominantly, Wolves ride **the taking risks rhythm.** In a University of Chicago study,[11] 172 men and women between twenty and forty were divided by chronotype and tested on the Domain-Specific Risk-Taking (DOSPERT) Scale to assess their comfort level with risk in five domains — ethical, financial, health/safety, recreational, and social. Subjects rated on a scale of one to seven (one being "extremely unlikely" and seven being "extremely likely") their reaction to situations like:

1. Admitting that your tastes are different from those of a friend.
2. Going camping in the wilderness.
3. Betting a day's income at the horse races.
4. Investing 10 percent of your annual income in a moderate-growth diversified fund.
5. Drinking heavily at a social function.
6. Taking some questionable deductions on your income tax return.
7. Disagreeing with an authority figure on a major issue.
8. Betting a day's income at a high-stakes poker game.
9. Having an affair with a married man/woman.
10. Passing off somebody else's work as your own.

The findings:

Male Lions were most likely to take social risks, like disagreeing with an authority figure or speaking their minds at a meeting.

Every other risk-taking prize went to the Wolves. Female Wolves were most likely to take ethical, financial, and health risks; male Wolves were most likely to take recreational risks.

Wolves are impulsive and novelty seeking, but they're also intelligent. Their best risks are calculated. Every team should have at least one Wolf to say, "Let's go for it!" They might make a lot of mistakes, but they're smart enough to learn from them.

The Rhythm Recap

The morning ritual rhythm: When Lions make good use of the early morning hours to chart a path to world domination.

The big idea rhythm: When all chronotypes use the power of daydreaming to come up with the next big idea that makes them rich.

The taking risks rhythm: When Wolves use their comfort level with wild chances and just might bulldoze their way to greatness.

THE WORST TIME TO GET RICH

12:00 p.m. to 2:00 p.m. At lunchtime, all chronotypes are alert and focused, and therefore **not** daydreaming or in risk-taking mode.

THE BEST TIME TO GET RICH

Dolphin: 9:00 a.m. to 12:00 p.m., sleep inertia daydreaming.

Lion: 6:00 a.m. to 9:00 a.m., when you've got the working world to yourself.

Bear: 7:00 a.m. to 9:00 a.m., sleep inertia daydreaming. **8:00 p.m. to 11:00 p.m.,** evening daydreaming

Wolf: 7:30 a.m. to 10:00 a.m., sleep inertia daydreaming. **9:00 p.m. to 12:00 a.m.,** evening daydreaming.

MAKE A DEAL

Failure: Negotiating for a car or a house or a raise when you're tired, hungry, or unreasonable, and winding up on the losing end of the bargain.

Success: Negotiating for a car, house, or raise when you're at peak concentration and energy, and winding up on the winning end of the bargain.

The Simple Science

By now, you already know that you're at your sharpest, strongest, and clearest during peak alertness hours.

- **Dolphin: Mid-afternoon**
- **Lion: Early morning**
- **Bear: Mid-morning**
- **Wolf: Early evening**

At those times, you will be able to make the necessary calculations and be most articulate in a negotiation. You also know that, regardless of the time of day, if you are sleep-deprived, you'll be irritable, foggy, and incapable of arguing for what you want to your greatest potential.

But negotiating for a good deal isn't *only* about being on top of your chronorhythmic game. Other factors have to be taken into account about your capabilities and those of the person you're negotiating with.

The ruthlessness rhythm is the time of day when ethical boundaries are blurred. Psychologists call it the morality effect. Anyone who goes into a car dealership knows that each side in the negotiation is going to do whatever he or she can to get a good price and package, and might say or do anything to get it. People are more (or less) likely to be unethical at certain times of the day. According to a study[12] by researchers at Cornell and Johns Hopkins, time of day determines when people are most likely to cheat on a game. Their subjects were asked to play games, given a financial incentive (higher scores = more money) to win, and told to self-report their scores. The morning types were more honest about their scores in the morning. The evening types were more honest in the evening. Neither type was more honest than the other, but each had its distinct moments of compromised morality.

Since every reader of this book is a fine, upstanding citizen who would never lie or cheat to get a better deal, simply figure out the chronotype of your opponent—say, a car dealer—and schedule your negotiations in accordance with his or her morality effect. Tired people have compromised morality. Alert people are more honest. It's just a matter of asking a few key questions early on in the negotiation:

1. "I couldn't sleep at all last night. Do you have that problem? Insomnia?" **If the dealer is a Dolphin, negotiate in the afternoon.**
2. "Let's do a test drive first thing. Are you available at 6:30 a.m.?" Non-Lion dealers will put you off until later. If he agrees to meet first thing in the morning and is energetic, you know he's a member of the pride. **If the dealer is a Lion, negotiate in the morning.**
3. "I get so damn sleepy after lunch. It's like someone drugged my soup. You?" If he wholeheartedly and sincerely agrees, he's probably a Bear (he likely is one anyway, just based on population percentages). **If the dealer is a Bear, negotiate in the mid-afternoon.**
4. "When do you close? I can't get to the dealership until 8:00 p.m." Lions will be dead tired by then, and Bears will do their best not to groan about a late night at work. A Wolf will pounce to see you at night. **If the dealer is a Wolf, negotiate at night.**

The on-your-game rhythm is determined by time of day, too. When are you in peak negotiation form? Short answer: Not before lunch.

Researchers[13] at Ben-Gurion University, in Beersheba, Israel, set out to determine if judicial rulings in parole hearings were based on facts alone or whether other issues—psychology, politics, social factors—came into play. As it turned out, none of those factors mattered, but the time a case was heard in relation to a judge's lunch and snack breaks made a significant difference.

In the morning session, judges started off with 65 percent favorable rulings, dropping to zero favorable rulings right before the lunch break, regardless of the circumstances of the hearing and the demographics of the prisoner seeking parole. After the lunch break, the positive rulings zipped back up to 65 percent. One could jump to the conclusion that hunger caused the judges to reject parole right before lunch. The researchers believe that the judges' mental capacity to make decisions was drained after a few hours of "choice overload." Subconsciously, the judges probably thought, "I've said yes too many times already, so I better start saying no."

Part of the pre-lunch dip in alertness is hormonal. Fasting hypoglycemia is commonly known as "low blood sugar," or what happens if you go longer than four hours without eating. Insulin, the hormone in charge of distributing glucose (energy) throughout your body, starts to work overtime, and the result is feeling confused, dizzy, tired, anxious, and irritable—that "I need to eat *now!*" feeling. When you are in that state, you can't focus on anything but food, and your ability to negotiate or even have a conversation will be impaired.

Even judges are susceptible to decision-making fatigue and hunger. Regular citizens can't be expected to close the deal when tired, hungry, or drained. So plan your negotiations at a time when you are well fed, well rested, and mentally fresh. If you're on your fifth negotiation that day, be aware that your opponent is likely to get the better of you. Experience is a valuable tool, but it's not as powerful as timing.

The Rhythm Recap

The ruthless rhythm: When you (or your opponent's) morality and ethics are compromised, making you (or him) more likely to lie and cheat.

The on-your-game rhythm: When it's been a while since you last ate, you feel drained and hungry, and you aren't able to concentrate on the negotiation.

THE WORST TIME TO MAKE A DEAL

Right before lunch, especially after a sleepless night. You will cave.

THE BEST TIME TO MAKE A DEAL

Dolphin: 2:00 p.m., after lunch and the postprandial afternoon dip in energy and attention.

Lion: 8:00 a.m., after breakfast.

Bear: 3:00 p.m., after lunch and the postprandial afternoon dip in energy and attention.

Wolf: 4:00 p.m., after lunch and the postprandial afternoon dip in energy and attention.

"I'M BETTER AT SAYING NO"

"As the supervisor of my department, I have to negotiate all day with vendors and with my own staff," said **Ben, the Bear.** "I'm not making big-money deals but doing small things, like saying when an employee can take a vacation and who gets what shift. They're always trying to get the best deal for themselves and I have to keep the entire department in mind. Anyway, the science about negotiating made me curious. I wanted to see if I caved more in the morning and less in the afternoon. The science doesn't lie. When I'm tired, I cave, especially if the person asking has a lot of energy. I can't counter a Lion asking for a favor early in the morning, or a Wolf hitting me up at the very end of the day. Saying yes too easily causes problems for me later on, when I have to scramble to fill spots. So I created a policy in the department: I'm only taking personal/vacation day requests and shift-switching requests between 3:30 and 4:00 p.m. This one change has made my job so much easier! I'm better organized, and I'm better at saying, 'No, that doesn't work for me.'"

SELL

Failure: Being unable to persuade anyone to buy whatever you happen to be selling.

Success: Projecting trustworthiness and bending just enough to coax your customer or client into buying your wares.

The Simple Science

The dance between a buyer and a seller is delicate. It might seem like you've succeeded in leading a client or customer to buy what you're selling, but then, abruptly, they decide to back out. It's not always easy to know what tips the scales in your favor. Some people are born salespeople and could sell snow to polar bears. But most of us could use some time-wise tricks to close the deal.

Perhaps the most important tool for a salesperson is to appear trustworthy. **The trustworthiness rhythm** is written all over your face. You don't have to be attractive to look honest. In fact, being attractive makes you seem *less* trustworthy. In a Princeton University study,[14] researchers digitally created eleven female faces they'd objectively ranked from "least average" to "most attractive," with "typical" faces in the middle. Using a nine-point scale, the study's participants were asked to rate the faces on trustworthiness. The results revealed that the least attractive and most attractive faces were not considered to be as trustworthy as the "typical" faces.

In an unrelated Swedish study,[15] researchers asked forty participants to rate photos of people's faces on their fatigue and sadness. Some of the people in the photos were, indeed, exhausted, having stayed awake for thirty-one hours before being photographed. The participants rated the sleep-deprived people as having hanging eyelids, red, swollen eyes, dark circles, more wrinkles, droopy mouth corners—not pretty. In fact, the fatigued people were described as "sad" by the participants. Sadness might be a typical emotion, but on the faces of salespeople, it's *not* a typical expression, and not a ringing endorsement of your trustworthiness. Do not attempt to sell when you are tired, or during the sleepy hours of the day.

Dolphins: I'm sorry to say, you probably don't have promising careers in sales.

Successful selling depends on not coming off as desperate to make the sale. During certain times of the day, your hormones will make you aggressive. From a buyer's perspective, your aggression is a turnoff. A Dutch study[16] found that higher levels of the aggression hormone testosterone reduced interpersonal trust. Using an MRI scanner, researchers tested twelve women to see how their brains reacted when they evaluated people's trustworthiness. The sole basis of the women's judgment was the facial appearance of the people they were evaluating. The subjects who'd been administered testosterone were less likely than the control group to judge a face as trustworthy. When that aggression hormone is flowing, a buyer is less likely to view a seller as safe. For the average Bear, testosterone peaks in the early morning and after exercising. When selling, know that your buyer is likely to be distrustful during those times.

A Japanese study[17] makes a similar connection between elevated cortisol levels and distrust. Cortisol has a definitive circadian rhythm, but it also goes up in stressful situations. If your buyer feels stressed out, he or she will react with distrust, and it's unlikely that you will close the sale.

The flexibility rhythm, or when your brain is at its most agile and able to see all the angles, will help you make a sale. A Mexican study[18] of eight subjects tested their cognitive flexibility every 100 minutes over a twenty-nine-hour period. Executive function declined sharply during off-peak alertness and performance times, which resulted in impaired problem-solving and decision-making skills. When you are attempting to make a sale and need to think fast on your feet to head off a potential buyer's hesitation and objections, your brain will be slower and less calculating during your chronotype's circadian downtimes.

However, you can remedy at least one daily downtime easily enough. According to a Belgian study,[19] a half-hour afternoon nap or the same amount of time exposed to bright light enabled subjects to overcome their afternoon dip in cognitive flexibility. Compared to the control groups, the subjects who napped and were exposed to bright light scored better on tests that measured their ability to switch from one subject to the next and to think about more than one thing at a time—exactly the kind of two-steps-ahead brainpower needed to make a sale. Since a nap isn't feasible for most office workers, exposure to bright light in the afternoon is a viable option for hitting the reset button on your selling powers.

The Rhythm Recap

The trustworthiness rhythm is when you put your best face forward to gain a potential buyers' trust.

The flexibility rhythm is when you can think fast and juggle more than one thought at time to stay one step ahead of the objections of your buyer.

THE WORST TIME TO SELL

Before 10:00 a.m. and after 10:00 p.m., when you're most likely to appear tired, which affects the perception of your trustworthiness. Also, you'll be off your cognitive flexibility peaks in the early mornings and late nights.

THE BEST TIME TO SELL

Dolphin: 5:00 p.m. to 9:00 p.m. Your cortisol is flowing, making you wide-awake and cognitively agile. Take advantage and sell!

Lion: 10:00 a.m. to 3:00 p.m. Quit before your cortisol level drops in the afternoon, and you start to feel (and look) tired.

Bear: 10:00 a.m. to 6:00 p.m. Bears make great salespeople and will be at their best from mid-morning to mid-evening. Take a break in the afternoon to reset your cognitive flexibility and you'll be good to go.

Wolf: 4:00 p.m. to 10:00 p.m. Take the closing shift, work the phones, or schedule meetings when your face appears wide-awake, your cortisol is pumping, and you're ready to dazzle.

CHAPTER 14

Fun

BINGE-WATCH YOUR SHOWS

Failure: Getting sucked into watching hour after hour of your favorite shows late at night, causing guilt, insomnia, and sleep inertia.

Success: Getting sucked into watching hour after hour of your favorite shows early in the day, and turning off the TV at a decent hour to avoid guilt and sleeplessness.

The Simple Science

By now, you have a breadth of knowledge about how circadian rhythms are affected by when you sleep, eat, exercise, work, think, and learn. How you have fun and spend your leisure time can disrupt or reinforce circadian rhythms as well.

Watching TV at night is as American as apple pie. The workday is over, the dinner dishes cleaned, the family plunks down on the couch to relax in front of the widescreen. My family watches TV, too, and we have our favorite shows as a unit and as individuals. On some nights, the TV goes on at 8:00 and stays on until the Power-Down Hour. Three hours of staring at the flickering screen — often, with phones and tablets on as well — isn't the worst health habit by far. But we all know that too much of anything isn't good for you, and that includes TV.

The first thing to consider is **the insomnia rhythm**, or how a TV's blue light emissions beaming into your eyes at night scrambles your sleep/wake cycle. According to your evolutionary bio-time, it should be pitch-black. Your pineal gland becomes confused. It's supposed to be night, and it should be secreting melatonin to make you feel sleepy, but all this sunlight-mimicking bright blue wavelength light must mean that it's still daytime, and that melatonin should be suppressed. Watching TV from ten feet away in an otherwise dark room won't disrupt your rhythms too badly. But anyone with chronic or sporadic insomnia will be affected by binge-watching on close-up devices like a tablet or phone.

For Dolphins, any bright light exposure after dark could set off an

upside-down hormonal cascade, keeping them awake for hours. It is interesting to me that many of my Dolphin patients think they need TV to fall asleep. In fact, it's keeping them locked in a sleep-deprivation cycle.

The depression rhythm associated with TV bingeing is a chicken-and-egg phenomenon. A University of Texas study[1] of 316 young adults linked the practice with loneliness, depression, impulsivity, and addictive behavior. Heavy TV users couldn't stop themselves from watching just one more episode, which turned into another, and another. But which came first? The depression and loneliness, or the marathon sessions of *Breaking Bad*? This is such a new psychological phenomenon that, at this writing, science doesn't know. We do know, however, that Wolves are more likely than other chronotypes to struggle with loneliness, depression, and addiction, and that it's only too easy for Wolves to stay up late watching TV, doing major damage to their already out-of-sync with social norms bio-time.

A Norwegian study[2] connected in-bed screen use (watching TV, computers, gaming consoles, hand-held devices) with insomnia and daytime sleepiness. The connection has been proven many times over. The Norwegians also found that, among their more than 500 subjects, the amount of in-bed screen time correlated with chronotype. Morning types used screens the least in bed. Evening types used them the most, which made their insomnia and morning grogginess symptoms worse.

For Wolves, watching more than two hours of nighttime TV will definitely keep you up at night, make you tired throughout the day, and exacerbate your most damaging personal behaviors.

The overeating rhythm is also associated with binge-watching TV. Glutting oneself with one indulgence opens the door to others. Science has drawn a clear map from your couch to the refrigerator. Researchers at Bowling Green State University did a study[3] with 116 overweight middle-aged adults in a weight loss program and found that binge-eating and binge-watching go hand in hand. If you watch TV, you will eat, even if you are actively trying to lose weight.

For Bears and Wolves, too much television will result in too many calories consumed. Watching TV (and therefore eating too much) at

night is the worst possible combination if you are trying to shed a few pounds.

Is binge-watching all bad? Of course not! Passing a Sunday afternoon with your favorite show and a bowl of popcorn is a fun experience and a cultural touchstone. I wouldn't miss bonding with my son over Barrett-Jackson car show marathons for the world. The time-wise trick is to do your TV bingeing during daylight hours, when the blue light won't keep you up and when overeating is unlikely to confuse your metabolism. Bingeing with friends and family is a way to *prevent* loneliness and depression. Wolves would do well to watch with someone who has the self-regulatory control to turn off the TV after two or three episodes, preventing a downward spiral into a seemingly harmless addiction that has serious emotional consequences.

The Rhythm Recap

The insomnia rhythm: When binge-watching (in or out of bed) interferes with the release of melatonin and causes sleep-onset problems.

The depression rhythm: When too much TV results in loneliness and depression (or the other way around), mainly in Wolves.

The overeating rhythm: When indulging in one kind of binge leads to another.

THE WORST TIME TO BINGE-WATCH YOUR SHOWS

After 10:00 p.m. All chronotypes with social-norm work schedules should turn off all screens by 10:00 p.m. to begin your nightly Power-Down Hour (see page 190). This includes hand-held devices.

THE BEST TIME TO BINGE-WATCH YOUR SHOWS

Dolphin: 10:00 a.m. to 2:00 p.m. A morning/afternoon stretch during sleep inertia. By afternoon, your laser clarity kicks in, and that's when the TV should go off.

Lion: 7:00 p.m. to 10:00 p.m., when you're winding down. Lions who have the occasional bout with insomnia should not watch in bed with hand-held devices.

Bear: 3:00 p.m. to 9:00 p.m. on weekends. Don't fall into the trap of staying up all night on Saturday. It'll wreck your sleep patterns for the upcoming workweek.

Wolf: 5:00 p.m. to 11:00 p.m. on weekends. Don't fall into the trap of staying up all night on Saturday. It'll wreck your sleep patterns for the upcoming workweek. Wolves should not TV binge alone. If you have a friend or family member watching with you, you'll be able to cut yourself off and stop the "one more episode" spiral.

THE DARK TRIAD

Great name for a hit TV show, right? Dark triad is actually a psychological term. It refers to three traits that, in the same person, make a human monster.

Psychopathology: remorseless, impulsive, callous.
Narcissism: prideful, egomaniacal, not empathetic.
Machiavellianism: manipulative, deceitful, exploitive, scheming.

Here's a very short version of a standard Dark Triad Personality Test.[4] Indicate if you agree or disagree with the following statements:

1. It's not wise to tell your secrets.
2. I like to use clever manipulations to get my way.
3. Whatever it takes, you must get the important people on your side.
4. Many group activities tend to be dull without me.
5. I like to get revenge on authorities.
6. Payback needs to be quick and nasty.

The more strongly you agree, the darker your personality.

The bad news for Wolves: Research[5] has found that, regardless of gender, you are most likely to have these traits.

The good news for all chronotypes: The most compelling and fascinating TV characters, the ones we love to binge-watch, exhibit the traits in spades. When we line up the TV characters' traits with their chronotypes, expect the villains to be night creatures.

Take the above test again, but this time, pretend that you are...Walter White of *Breaking Bad*, Don Draper of *Mad Men*, Cersei Lannister of *Game of Thrones*, or Frank Underwood of *House of Cards*.

See what I mean? Dark triad characters are evil, selfish, devious, manipulative, unscrupulous—and compulsively watchable.

LOG ON

Failure: Compulsively surfing the Internet throughout the day and/or late at night, causing insomnia, next-day grogginess, lack of productivity, and anxiety.

Success: Surfing the Internet for fun and information during downtimes, and logging off an hour before bed to prevent insomnia.

The Simple Science

Before I get into it, I want to draw a line between what I call "fun and functional" and "time- and brain-sucking" Internet use.

"Fun and functional" Internet use is going online with a vague purpose, shopping, sending messages, staying informed, and keeping in touch on social media for an hour or so per day.

"Time- and brain-sucking" (TABS) Internet use is everything else, the endless Googling about things you don't really care about, mindlessly clicking from one link to the next, checking social media platforms constantly, compulsively.

Although you are probably aware that TABS Internet use is bad for your work, you might not be aware of what it's doing to your bio-time and general happiness.

As you know, each chronotype has several hours of on-peak alertness each day. Focus is sharp, and energy is up. All that concentration and arousal are wasted on time- and brain-sucking Internet use. I have fallen into this particular black hole myself on occasion. It starts well, with purposeful Googling, and then I look up three hours later and have nothing to show for it. Letting your clicking finger take you to unexpected places has some value. During easily distractible, creative hours, you might find a pot of gold at the end of the random Internet thread rainbow. But during on-peak hours, let that thread go, or squander your daily **productivity rhythm.**

Social media is the enemy of knuckling down. Guess which chronotype is most vulnerable to the siren call of Facebook? According to a Spanish study[6] about the degree to which Facebook intrudes on your life, evening types won (lost?). A Japanese study[7] confirmed the Wolfish social media obsession among its nearly 200 medical school student subjects. Of particular note, female evening types can't get enough of Twitter. It's all about impulsiveness. Wolves see the notification pop-up on their home screen, and they just can't resist swiping to see what's going on, and then they get sucked in.

They do so at the peril of their neural pathways. **The screw-up rhythm,** or when you're dazed and confused and prone to making mistakes, correlates with how often you log on. A British study[8] of 210 subjects aged eighteen to sixty-five linked Internet use with memory and motor-function failure. The more time subjects spent on their smartphones, the more mistakes they made in normal everyday logistics. Their "cognitive failures" were things like missing appointments, zoning out when spoken to, forgetting objects and places, and misreading street signs.

Apparently smartphones can make you stupid.

Since most of us carry our phones in our pockets or handbags and check them constantly, it's kind of hard to pin down exactly when time- and brain-sucking Internet use is most destructive. **The traffic rhythm,** or when most people go online, *is* well known, and works against certain chronotypes. Internet rush hour—the time of day when traffic on the information superhighway is heaviest—is from 7:00 p.m. to 11:00 p.m. For Lions and Bears, this is actually a good thing, sort of. Bears are off-peak at that time, and if they can log on after dinner to check their favorite blogs and sites and log off by 10:00 p.m., Internet use won't affect their bio-time. Same for Lions.

The traffic rhythm is bad news for Dolphins, though, since *any* exposure to brain-arousing activity—like clicking on an engaging article or commenting on a friend's intriguing post—will tip their topsy-turvy hormonal and circulatory systems heavily in the wrong direction.

Wolves face the steepest challenges. I mentioned a Norwegian study[9] that proved that using electronic devices in bed—a predominantly Wolfish practice — causes insomnia and next-day grogginess. I get it: It's hard to shut down your tablet at night. But if you don't, it'll throw your bio-time out of sync for days. If you think it's hard to log off at night, that goes double for teens, a subgroup of the population that tend to be predominantly Wolves. In a Taiwanese study[10] of nearly 3,000 incoming college students, researchers divided subjects by chronotype and tested for a variety of personality traits and behaviors, including obsessive-compulsive disorder (OCD), anxiety, and disrupted sleep habits. Along with being more compulsive and Internet-addicted, the study's teen Wolves relied on weekends to catch up on sleep (which does them no favors) and were more anxious than Lions and Bears. The bright side? The more family support they received, particularly from their mothers, the better off they were in terms of anxiety, OCD, and Internet addiction. I'm not advocating helicopter parenting, but there is nothing wrong with telling a kid, "Shut that thing off already!"

In fact, there's nothing wrong with telling yourself the same thing. Most chronotypes will be successful at avoiding the Internet black hole earlier in the day. **The willpower rhythm** was explored by social psychologist Roy Baumeister, who in 1997 discovered the phenomenon of "ego depletion" or "willpower depletion." In a famous study,[11] Baumeister's subjects sat down at a table with a plate of freshly baked cookies and a bowl of radishes. Half the subjects were instructed to enjoy the cookies; half were offered the less appealing radishes. Then all of them were asked to solve a challenging math puzzle. The radish-eaters quit after eight minutes. The cookie monsters kept at the puzzle for nineteen minutes. The researchers concluded that, by exhausting their willpower by resisting the delicious cookies, the radish group depleted their energy for puzzle solving, and, theoretically, any other challenge they might face later on.

I described this study to a friend of mine, and her response was, "Don't want to deplete your willpower? Then eat the damn cookie!" Which may or may not defeat the purpose of resisting temptation in the first place. Does eating the cookie today make it easier to resist tonight? Something to think about for future study.

What does this study have to do with Internet self-control? It's all about resisting the cookies, be they the gooey chocolate variety or the cookies saved automatically on your web browser that take you to your favorite sites faster than a single keystroke. Since willpower is depleted over the course of the day, it's harder to log off at night. The willpower and productivity rhythms dovetail nicely for Lions and Bears. Dolphins and Wolves, not so much. Although their willpower is strongest in the morning, their productivity increases as the day goes on.

Ironically, technology has rushed in to solve the problem of excessive use of technology. As the saying goes, "There's an app for that." I wrote this chapter while using the Freedom app to block all web browsing for a three-hour chunk. Other blocker apps: Anti-Social, SelfControl, and Cold Turkey. Spending half an hour browsing for and then installing a web-restricting app is a time-wise investment that will pay off in a big way.

The Rhythm Recap

The productivity rhythm: When you are most productive during the day and shouldn't be wasting time online.

The screw-up rhythm: When excessive smartphone use makes you more likely to make stupid mistakes.

The traffic rhythm: When most people go online, clogging traffic on the information superhighway.

The willpower rhythm: When you have the ability to resist the lure of the Internet.

THE WORST TIME TO LOG ON

On-peak productivity hours, depending on chronotype; within an hour before bedtime. By now, you should know your on-peak times and bedtime. Log off one hour before sleep so you don't disrupt the flow of nighttime melatonin.

THE BEST TIME TO LOG ON

Dolphin: 9:00 a.m. to 3:00 p.m. Block browsing until **9:00 p.m.** Log off by **10:30 p.m.**

Lion: 6:30 a.m. to 6:00 p.m. Log off by **9:00 p.m.**

Bear: 8:00 to 11:00 a.m. Block all but work-related sites until **7:00 p.m.** Log off by **10:00 p.m.**

Wolf: 9:00 a.m. to 3:00 p.m. Block browsing until **10:00 p.m.** Log off by **11:00 p.m.**

THE BEST TIME TO TWEET, POST ON FACEBOOK, AND ONLINE-DATE

Twitter: According to a two-year study[12] by sociologists at Cornell University of 509 million tweets, Twitter is at its most upbeat and enthusiastic at 8:00 a.m. to 9:00 a.m. on weekdays and 9:30 a.m. to 10:30 a.m. on the weekends. But if you want to track the latest Twitter war, log on at 10:00 p.m. to 11:00 p.m., when tweets are emotionally charged and users are fully engaged.

Facebook: Prime time is 7:00 p.m. to 8:00 p.m. (including weekends), per a 2015 study[13] by Klout and Lithium Technologies of over one hundred million posts. Post then to get the most shares, comments, and likes.

Online dating sites: According to two major dating sites (match.com and plentyoffish.com), the most popular time of year to sign up is between New Year's Day and Valentine's Day. The most user-active time to log on? Evenings at 8:00 p.m.

PLAY GAMES

Failure: Playing cards, computer games, or board games when you're more likely to cheat, lose, or wind up with insomnia.

Success: Playing cards, computer games, or board games when you're more likely to be a good sport and win.

The Simple Science

Although most of us play games for fun, to kill time on a subway or bus, or to bond with friends and family, some people play games to win, and they'll do anything for that outcome.

The cheating rhythm is the time of day you're more likely to bend the rules. Say you're playing Monopoly over Thanksgiving weekend with your brother-in-law, a man who equates his personal self-worth with building a house on Park Place. You can anticipate when he'll steal money from the bank (more than he usually does) based on his chronotype. For a study[14] by researchers at Cornell and Johns Hopkins, subjects played games—matrix problems or rolling-the-dice games—with a financial incentive (the better they did, the more money they would receive for participating) and, here's the trick, they were told to record their own scores. The morning-oriented subjects were more honest about their scores in the morning and exaggerated their scores in the evening. The evening types tended to be more honest in the evening and to exaggerate their scores in the morning. One type was not more honest than the other, but each had ethically shaky moments.

Is your blowhard brother-in-law a Lion? If so, he'll probably cheat at night. If he's a Wolf, he'll cheat in the morning and afternoon.

If he's a Bear, he's more likely to cheat in the evening, when he's tired.

Your Dolphin sister-in-law will be vaguely off-peak all morning and into the afternoon. But, since insomniacs tend to be nonaggressive and overly cautious, she's unlikely to cheat. She might be annoyingly officious about everyone else's conduct, though.

The insight rhythm pertains to when your brain is best able to be creative and connect seemingly unrelated dots to form a clear picture. Examples of games that take insight: charades, Pictionary, crossword puzzles, anything that requires you to think and have an aha! moment to get to the next level or solve the puzzle. You'll do better to play these games when you're a little tired, a little unfocused, easily distracted—in other words, at non-optimal times. But watch out! During those same times, you're also more likely to cheat.

The strategy rhythm, the flip side of the insight rhythm, is when your brain is best able to analyze and use logic to stay one step ahead. Examples of games that involve strategy: dominoes, chess, checkers, backgammon, Scrabble, and most card games (especially poker and bridge) that require you to do the math—to think with cold, calculated logic. You'll do better playing strategy games when you're wide-awake, able to concentrate, and laser-focused—in other words, at optimal times.

The chance rhythm is when your brain is best able to take calculated risks and you have the self-control to collect your chips and walk away before you bet your mortgage payment. Examples of games of chance: blackjack, roulette, Go Fish, any game that is completely dependent on luck. (One could argue that blackjack is a game of strategy, but most of us rely on standard rules and pray for an ace.) Most games of chance aren't won or lost on bio-time. You could be a Bear, playing a cutthroat Lion at Go Fish on-peak, off-peak—doesn't matter. The winner is determined by fate, not circadian rhythm.

However, if you are playing games of chance in a casino with real money, the best strategy is keeping a close eye on your watch. You won't find clocks in any casino for this very reason: When it's late at night, and Bears and Lions are off-peak, they're more likely to throw caution to the wind and let it ride. Why? When you're tired, you become irresponsible.[15] Sleep deprivation—like staying up until all hours at the craps table—impairs decision making. When you're fatigued, the prefrontal cortex, the brain region that controls higher-order processes like judgment ("Is this a good idea?" "Is this a bad idea?"), switches off. A study[16] by the Walter Reed Army Institute of Research put this to the test. Thirty-four participants started playing gambling games on a computer. They learned as they played that certain decks yielded reliable wins, some reliable losses. For a while, the participants made smart decisions about which decks to play. But as time wore on, the subjects started to make bad decisions—and started losing. Sleep deprivation compounded uncertainty, that feeling that you don't know how things will work out. When you're tired and up to your neck in uncertainty, you will take bigger risks that are unlikely to pay off.

Wolves are risk-takers, especially when it comes to gambling. According to a Duke University study,[17] 212 subjects ranked their risk-taking in five different areas, including financial, and they found that Lions held off on taking money risks, especially those related to gambling, and that Wolves were most likely to go for it. If you are a Wolf and you are planning a trip to Las Vegas, please take a Lion with you to tear you away from the tables before you start to get reckless.

The Rhythm Recap

The cheating rhythm: When ethical lines are blurred and you're likely to cheat.

The insight rhythm: When your brain is creative and able to connect random dots solving puzzles and playing games that require insight.

The strategy rhythm: When your brain is most analytical and focused, and best able to play games involving strategy.

The chance rhythm: When your brain is best able to figure out calculated risks while playing games that depend on luck.

THE WORST TIME TO PLAY GAMES

2:00 a.m., even for Wolves. When you're tired, you're likely to cheat, have less fun, make poor judgments, and lose your shirt.

THE BEST TIME TO PLAY GAMES

Dolphin: Games that require insight, **10:00 a.m. to 2:00 p.m.** Games of chance and games that involve strategy, **4:00 p.m. to 10:00 p.m.**

Lion: Insight, **5:00 p.m. to 9:00 p.m.** Chance/strategy, **7:00 a.m. to 3:00 p.m.**

Bear: Insight, **6:00 p.m. to 10:00 p.m.** Chance/strategy, **10:00 a.m. to 2:00 p.m.**

Wolf: Insight, **8:00 a.m. to 2:00 p.m.** Chance/strategy, **4:00 p.m. to 11:00 p.m.**

ADDICTED TO LEGEND OF ZELDA?

According to the *Diagnostic and Statistical Manual of Mental Disorders,* Fifth Edition, Internet gaming addiction (IGA) is diagnosed by the following criteria:

- Preoccupation or obsession with Internet games.
- Withdrawal symptoms when not playing them.
- A buildup of tolerance (more time needs to be spent playing).
- The person has tried and failed to stop or cut down on time spent playing.
- The person has lost interest in other life activities.
- The person has continued to overuse Internet games even when he knows how much playing affects his life.
- The person has lied about how much time he spends playing.
- The person uses playing as a way to escape real life.
- The person has lost or put at risk opportunities or relationships because of Internet games.

Does this sound like you or someone you know? Gaming addiction is a new disorder, but one that can't be underestimated, in part because it affects the young. According to a Turkish study[18] of 741 adolescents, IGA is predictable by chronotype, personality traits, and gender. In a nutshell, introverted, disagreeable Wolf boys are most likely to become addicts. Extroverted, agreeable Lion girls are the least likely to become addicted.

To better understand why certain types become addicted to Internet games, you have to ask what they're getting out of it. The short answer is dopamine. When you reach the next level, a release of that happy hormone lights up the reward center of your brain. It's not dissimilar to taking a drug. You aren't addicted to Zelda or cocaine. You're addicted to dopamine.

Dopamine runs on bio-time. It and melatonin have an inverse release schedule. The pineal gland releases melatonin at night, as you know, to make you sleepy. Researchers have discovered[19] that the pineal gland has receptors for dopamine. When dopamine is released and the pineal receptors catch it, the gland switches off melatonin in the morning so you can wake up. There is a multidimensional link between screen use and insomnia, and we now know that dopamine plays an active part.

The usual detox prescription for IGA is to find ways in your real life to get a dopamine rush. A time-wise trick to facilitate this is to use a web blocker (page 278) to go cold turkey on games during off-peak times, when all types are most vulnerable to impulsivity. Whenever you are stressed and feel a psychological need to escape, instead of turning to a game, do dopamine-boosting activities such as Go for a Run (page 121), Practice Yoga (page 128), Meditate (page 152), Play Music (page 259), and Read for Pleasure (page 307).

READ FOR PLEASURE

Failure: Only reading short pieces or not reading at all.

Success: Cracking open a good book to ignite memory and imagination and blaze new neural pathways.

The Simple Science

Reading at any time of day or evening is good for you. It superpowers your brain and blazes neural pathways that boost memory, creativity, vocabulary, productivity, and empathy.[20] Reading keeps you informed about what's going on in the world. It slows brain deterioration[21] due to Alzheimer's and dementia, and it makes you a more empathetic, well-rounded person.[22] Reading is to the mind what yoga is to the body: It keeps you strong, flexible, and open to new ideas and perspectives.

Technical reading for work should be done at on-peak alertness times for better acquisition. Reading for pleasure? You will not hear me say, ever, that there is a bad time for that. In my capacity as a doctor and researcher, I read all day long, and then, when I have leisure time, I read some more. Reading is a good addiction to have, one that feeds on itself. If you read one book and enjoy it, you are likely to read another, and another.

Dolphins, take note: Reading lowers your cortisol level, which will help you before bed. British neuropsychologist David Lewis of the University of Sussex researched **the calming rhythm** in 2009. He had volunteers exercise and do mental tasks to raise their stress level, and then tried various methods to calm down, such as listening to music, having a cup of tea (it was an English study, after all), playing video games, taking a walk, and reading a book. All of the strategies worked to lower stress, but the hands-down winner for slowing heart rate, reducing muscle tension, and reducing stress was reading. After only six minutes with a book, subjects' stress was down 68 percent. Discussing his results with the *Telegraph*, Lewis said, "It really doesn't matter what book you read, by losing yourself in a thoroughly engrossing book, you can escape from the worries and stresses of the everyday world and spend a while exploring the domain of the author's imagination. This is more than merely a distraction but an active engagement of the imagination as the words on the printed page stimulate your creativity and cause you to enter what is essentially an altered state of consciousness."

I urge you all to create some kind of **ritual rhythm**, to establish a set time every day to soak up the benefits of reading. You might choose to read during your commute or at lunch hour. If I had to recommend the most practical and beneficial time to read, I'd say the Power-Down Hour (page 190) before bed. It's what many of us already do, and it might be the only time all day when you can sit down with a good book. A couple of caveats:

1. **If you struggle with insomnia at all, don't read in bed.** Read in a cozy chair near your bed, or even while reclining on a couch. To overcome insomnia, you have to associate bed with sleep (and sex) only.

2. **Ebook readers might cause problems.** This has become a controversial topic as more and more people use and love ereaders. A close friend of mine doesn't go anywhere without her Kindle and will probably be buried with it. A Wolf, she was dismayed when I shared the findings of a new Harvard Medical School study:[23] reading an ebook in the hour before bed delayed sleep more than reading a print book under normal lamplight, and it also increased sleep inertia the next day. The culprit was ereaders' emission of short-wavelength enriched light and the melatonin suppression it causes. There is technology out there to block or filter out blue wavelength light — shields for devices, glasses, white-light bulbs. For more information and my recommendations about products, go to www.thepowerofwhen.com.

The Rhythm Recap

The calming rhythm: When cortisol level and stress can be reduced naturally if you read a book for as little as six minutes.

The ritual rhythm: When you create a set time for positive, healthful activity like reading to ensure that you do it every day.

THE WORST TIME TO READ FOR PLEASURE

No. Bad. Time. To. Read.

THE BEST TIME TO READ FOR PLEASURE

Open your mind and eyes to the written word every day for multidimensional benefits to the mind and spirit. I recommend reading during your nightly Power-Down Hour in preparation for sleep.

Dolphin: 10:00 p.m.

Lion: 9:00 p.m.

Bear: 10:00 p.m.

Wolf: 11:00 p.m.

TELL A JOKE

Failure: Fumbling the punch line or trying to be funny when your audience is in a serious mood.

Success: Nailing the punch line and being funny when your audience is primed to laugh.

The Simple Science

Ever wonder why comedy is associated with nighttime? After-hours TV comedy shows often have the words "late" or "night" in the title — *Saturday Night Live, The Tonight Show, The Late Show, Late Night,* etc. Comedy clubs don't open until dark. The comedy clubs I've been to artificially create a middle-of-the-night feel with dim lighting and either no windows or curtained windows. In the world of comedy, everything is funnier after midnight. Why? Humor is on bio-time, too.

In part, it's due to **the hormonal rhythm.** After dark, levels of serotonin — the happy, one could say "laffy," hormone — rise to calm you down as you make the transition to sleep. Meanwhile, the fight-or-flight hormone cortisol levels plummet. During daytime, the hunting hours, you are ramped up by cortisol. Humor is beside the point. But at night, under the influence of happy hormones, you are loose and relaxed — in just the right mood to laugh.

The intoxication rhythm makes it possible for you to "get" the joke. As I've explained, creative connections occur when focus is fuzzy. Your brain makes remote associations then, connecting random ideas in ways that might not make sense during peak alertness and arousal times. Comedy is about surprise — the unexpected punch line, anarchistic mayhem, taking something familiar and turning it upside down. Monty Python knew this and used the catchphrase "And now for something completely different" on their TV series. A good reason for the two-drink minimum in comedy clubs: When you are slightly impaired, you're neurologically primed to appreciate and understand the surprise of humor.

Intoxication comes in many forms. Alcohol. Marijuana. Sleep deprivation. I'm sure many readers have already observed this phenomenon in their own lives. Try it at home. Watch a *Three Stooges* movie stone-cold sober at 10:00 a.m., and watch it again at 10:00 p.m. after a couple of cocktails or glasses of wine. Big difference.

Too much impairment, though, is not so funny. If you've had a whole bottle of wine or stayed up for an entire night, the joke will be lost on you. Scientists at Walter Reed conducted a study[24] of fifty-four healthy adults, keeping them awake for forty-nine hours and showing them cartoons and witty newspaper headlines to study how severe sleep deprivation affected sense of humor. Not too surprisingly, *nothing* was funny to the study subjects after they had stayed awake for two days—even for those who were given coffee and stimulant medication to counter fatigue. Their prefrontal cortices (the region that controls decision making, judgment, sorting—the cognitive abilities you need to figure out whether a joke is funny or not) were shot.

The two rhythms above are about appreciating humor. The best time to hear a joke is late at night. But when is the best time to tell one? **The delivery rhythm** explains why Wolves tend to have a better sense of humor[25] than do other chronotypes. As we all know, nothing is quite as unfunny as a drunk person attempting to tell a joke in a rambling, incoherent way. They forget the punch line or have to go back and correct the badly described setup, as in, "A rabbi, a priest, and a monkey walk into a bar . . . wait, was it a rabbi, a priest, and a horse? . . . not a bar, I think it was a prison? . . ." The audience has fallen asleep while this jokester fumbles the lines. Telling a clear, concise joke requires sharpness and focus, which Wolves have in abundance late at night, when most people are off-peak and therefore most likely to appreciate humor.

Or perhaps Wolves have a better sense of humor due to their personality traits. As famous Wolf humorist Mark Twain once said, "The secret source of humor itself is not joy but sorrow. There is no humor in heaven." Evening types are more likely to suffer from anxiety, depression, addiction, isolation—hilarious stuff. If comedy indeed comes from pain, it's no wonder Wolves keep the rest of us laughing deep into the night.

The Rhythm Recap

The hormonal rhythm: When various hormones ebb and flow, affecting mood, alertness, and sleepiness—and the ability to get the joke.

The intoxication rhythm: When your brain is off-peak due to circadian fluctuation, sleep deprivation, or substance use, and jokes become funnier.

The delivery rhythm: When the joke teller has on-peak sharpness and focus for articulation and communicating well.

THE WORST TIME TO TELL A JOKE

6:00 a.m. to 9:00 a.m. Even if you are sharp and focused and can kill the delivery, people are not in the mood to giggle when they first wake up.

THE BEST TIME TO TELL A JOKE

Dolphin: 7:00 p.m. You have near-peak focus at dinnertime, when Bears and Lions are off-peak enough to appreciate the joke.

Lion: 2:00 p.m. Your focus is sharp enough to kill during the afternoon dip, when everyone else is lulling/LOLing.

Bear: 5:00 p.m. You hit a second wind in the evening and can kill it in meetings with Lions, who are just winding down, and Wolves, who are just waking up.

Wolf: 10:00 p.m. When the sun goes down and other chronotypes are off-peak, you hit your sharpness peak to kill and kill again.

TRAVEL

Failure: Feeling miserable, irritable, clumsy, stupid, slow, and exhausted for days on end after crossing multiple time zones.

Success: Feeling the modest effects of crossing multiple time zones for forty-eight hours or less.

The Simple Science

I've talked a lot about social jet lag—the irritability, grogginess, and fatigue associated with chronic chrono-misalignment. Remember, chrono-misalignment is when your circadian rhythm is out of sync with social norms that dictate the "when" of everything: when to sleep, eat, work, play, relax. Social norms happen to be Bear-friendly, but even Bears have to deal with chrono-misalignment when they stay up late on weekend nights, sleep in on weekends, eat at irregular times, or stare at screens after dark. Being socially jet-lagged by only an hour or two can cause significant problems and prevent you from being all you can be.

Another phrase about being out of sync with bio-time you need to know: "forced circadian desynchrony," or forcibly undoing your chronorhythm to an extreme degree. One cause is late-shift work. A far more common cause is air travel across time zones, or **the desynchrony rhythm** of jet lag.

Jet lag affects each chronotype differently.

- **Dolphins suffer terribly.** About half of the insomniacs in my practice can't sleep on planes due to hypersensitivity to the environment—the lights, noise, upright seats, people, food. They arrive at their destination completely stressed out and exhausted, which compounds sleeplessness even if they're staying in luxurious hotel rooms. If you are a Dolphin or have traveled with one, you know that the first two or three days in another time zone is pure misery. **Recommendations for Dolphins in flight: (1) On red-eye flights, you might benefit from taking a sleeping pill.**[26] **(2) If possible, schedule flights, even long ones, during daylight**

hours, when you aren't expected to sleep. You will lose a day in the air, but you'll feel better when you land. It's a trade-off.

- **Lions suffer traveling west but do better traveling east.** A Lion from Los Angeles, for example, benefits from flying to New York. For the first few days, his wake-up time fits a Bearish schedule. But look out, Lions, when traveling west. A British Lion traveling to New York will wake up at 2:00 a.m.

- **Wolves suffer traveling east but do better traveling west.** A New Yorker Wolf will instantly adjust to a Los Angeles schedule, but if she flies to Paris, she'll be unable to wake up until noon and won't feel tired until the middle of the night (maybe that's not such a bad thing, given Parisian nightlife).

- **Bears suffer equally flying east or west,** but not as severely as do the other chronotypes... *unless they drink alcohol.* The issue is dehydration. The plane's dry air and the overly salty food will suck the life out of you as it is. If you have a drink, dehydration will only get worse. Affable, easygoing Bears might get too comfortable in their seats and forget to move every hour or so. Immobility causes deep vein thrombosis and swelling. Have you ever taken your shoes off at the beginning of a flight and found it hard to get them back on at the end? Prevent it by getting your blood moving every hour in-flight just by walking up and down the aisle for a minute or two.

Traveling for a family vacation to an exciting foreign city or a tropical island is, for me, the very definition of fun. I spend many hours each week traveling for work as well, which isn't nearly as fun, but I take it in stride because I know the tricks to adjusting quickly. I've helped many high-flying patients recover from chronic jet lag. One is a man who commutes from New York City to Australia every month and had been jet-lagged for ten years. He's much better now because he follows **the resynchrony rhythm**, or getting back on bio-time in a new time zone.

I've worked out most of the time-wise tricks below in my own practice and have enhanced my own data with information from NASA guidelines[27] for multiple time zone travel for pilots. If the recommendations below are good enough for NASA cosmonauts going to the International Space Station, they're certainly good enough for anyone of any chronotype.

Note: The guidelines below are for traveling at least three time zones. If you're traveling one or two time zones, you'll need to follow the strategies for day one only or not at all. The human body needs one day per time zone traveling to adapt.

Traveling East, or Phase Advance (Waking Earlier; Going to Bed Earlier)

- **Day of flight:** No caffeine at all. Adjust your watch to your new time zone.

- **During flight:** After two hours on the plane, attempt to sleep for the remainder of the flight. Use the complimentary eye mask and earplugs, or bring your own. If you can't sleep, avoid light and/or wear sunglasses.

- **Upon landing** at your destination: Put on shades if they're not already on.

- **Day one** at your new destination: Wear sunglasses until 12:00 p.m. After 12:00, take your sunglasses off, and get as much direct sunlight as possible, especially between 1:30 p.m. and 4:30 p.m. If you're stuck inside all afternoon, take sunshine breaks for ten minutes each hour. You can have caffeine upon arrival, but no later than 3:00 p.m. Eat breakfast, lunch, and dinner (page 194) on your new schedule, even if you're not hungry. Exercise in the afternoon, preferably outdoors. Naps: no! A sleep aid can help you sleep later the first night.[28] NASA recommends using one, too. Don't bother setting an alarm. Sleep in as long as you can.

- **Day two**: Put on your sunglasses upon waking and keep them on until 10:00. After 10:00, take the sunglasses off and get as much direct sunlight as possible, especially from 11:30 to 2:30. If you're stuck inside, take sunshine breaks every hour. Caffeine: yes, but no later than 3:00 p.m. Eat on your new schedule, even if you're not hungry. Exercise in the afternoon, preferably outdoors. Naps: no!

- **Day three**: You'll feel normal by this morning, but continue wearing sunglasses before 9:00 and getting direct sunlight after 9:00 on an hourly basis.

- **Day four:** Congrats! You are now comfortably on a Bear's chrono-rhythm at your new time zone.

Traveling West, or Phase Delay (Waking Up Later; Going to Bed Later)

- **Day of flight:** No caffeine before the flight. Set your watch for your destination time zone. Wear sunglasses all day until the flight.
- **During flight:** As soon as you get comfortable, put on an eye mask and ear buds and listen to a relaxation audio program (go to www .thepowerofwhen.com for a download), and attempt to sleep. If the flight is long enough, use a sleep medication.[29] No caffeine for the duration of the flight. Wear sunglasses until the last two hours of the flight. Then take them off and get as much sunlight as possible through the plane window or artificial light with close-up screen exposure.
- **Day one** at your new destination: The sunglasses off, get as much direct sunlight as possible, especially in the evening. Use screens at night until bedtime. No caffeine after 6:00, and no naps. Exercise before noon, and eat on your new schedule, even if you're not hungry.
- **Day two:** Get as much direct sunlight as possible, morning through evening. No caffeine after 3:00. Exercise in the morning, and eat on your new schedule. If you're really not hungry, have something light, like a smoothie. Remember, eating on a schedule will help shift your biorhythm.
- **Day three:** Congrats! You are now on a Bear's chronorhythm!

The Rhythm Recap

The desynchrony rhythm: When you travel to a new time zone and suffer jet lag symptoms such as irritability, clumsiness, mental fog, and exhaustion.

The resynchrony rhythm: When you use specific strategies to rapidly resync your bio-time to your new time zone.

THE WORST TIME TO TRAVEL

When drunk. One drink in the air is more intoxicating than one drink on the ground (due to in-flight dehydration).

THE BEST TIME TO TRAVEL

For flights that take you three time zones or more from home:

Dolphin: Daylight hours, to save yourself a night of plane-related insomnia.

Lion: Late evening. For an overnight flight, you'll do slightly better to arrive very early in the morning.

Bear: Overnight, at your flight schedule convenience.

Wolf: Midnight. Take the last plane out, and you'll be able to sleep better during the flight.

"CURING JET LAG THROUGH THE STOMACH"

"One of my life goals is to travel to every continent and hike the mountains of the world," said **Robert, the Lion.** "But jet lag has ruined a lot of trips for me. Taking three or four days to recover on a five-day trip is a waste of time and money. I recently took a dream vacation to Hawaii—a five-hour time difference from Boston, going west. I followed the NASA/Breus plan to the letter, and I felt a lot better after two days and was fully functional on day three of a weeklong trip. Getting sunlight wasn't an issue, because I was outside all day (no shades). What made the difference for me was forcing myself to eat on the new schedule, even if I wasn't hungry. The brain clock and the stomach clock have to be synchronized to shift your rhythm. Sunlight changes the brain clock, and eating within a half hour of waking keeps the stomach clock on bio-time. My advice to travelers: The quickest way to cure jet lag is through the stomach."

PART THREE

THE POWER OF WHEN FOR LIFE

CHAPTER 15

Chrono-Seasonality

Until now, I've explained the ups and downs and ebbs and flows of how everything about you, from your alertness and mood to your creativity, changes over the course of a twenty-four-hour period. Circadian means "about a day," after all. However, your day-to-day bio-time does shift slightly over the course of a month, a season, and a year. The severity of those changes depends on your chronotype.

Does one chronotype suffer most profoundly from premenstrual symptoms?
Is one chronotype thrown for a loop during daylight saving time?
Which chronotype is especially susceptible to the winter blues?

In this chapter, I'll widen the lens and explain how the power of when applies to monthly, seasonal, and annual shifts, and how you can adjust your chronorhythm accordingly.

THE LUNAR RHYTHM

The moon's gravitational pull exerts a force on the movement of oceans, and it affects human bodies as well. Recently, the moon's phases have been linked to hormonal fluctuations inside each one of us. Researchers at the University of Basel in Switzerland conducted a study[1] of thirty-three men and women between the ages of twenty and seventy-four. The subjects slept in a lab, and their light exposure (artificial and natural) and melatonin levels were closely monitored. **In the days leading up to the full moon, melatonin started dropping dramatically, hitting its lowest level on the night of the full moon.** Melatonin rose to its highest level at day fourteen or fifteen of the twenty-nine-day lunar cycle.

The subjects' sleep duration, sleep quality, sleep depth, and ability to fall asleep were at their lowest at the full moon. Deep slow wave delta sleep decreased by 30 percent during that phase. Conversely, midway through the lunar cycle, the subjects slept more deeply, took less time to fall asleep, and slept for a longer time. Full moons make people feel restless. Folklore figured that out decades before Swiss scientists could prove it in a lab.

The chronotype most negatively affected by the lunar rhythm:

Dolphins. (Wolves get a break for once.) The Swiss scientists determined that, on average, subjects lost twenty minutes of sleep on the days before and during the full moon. Lions, Bears, and Wolves can handle a few days of shortened sleep. But, for Dolphins, several nights of diminished sleep quality and duration can set off a chain reaction of anxiety and insomnia that lasts for weeks. My recommendation for Dolphins is to be aware of the lunar cycle and use supplemental melatonin — 0.5 mg taken ninety minutes before your calculated bedtime — to counter the dips.

THE MENSTRUAL RHYTHM

Menstruation has its own rhythm: First, the follicular phase, when an egg comes to maturity in the ovarian follicle. Next, the ovulation phase, when the egg is released from the follicle. After that, the luteal phase that ends with menstruation. The whole process, from follicular phase to menstruation, takes approximately twenty-eight days.

Most ovulating women can attest to the mood and metabolic shifts over the course of the menstrual cycle. During the follicular phase, women tend to feel normal. During ovulation, sleep quality is compromised. During the luteal phase, the benchmarks of bio-time — body temperature, melatonin secretions, cortisol release, the quantity of REM sleep—are all affected,[2] and not, I'm sorry to say, in a good way. For most women, the hormonal changes cause increased stress and appetite (melatonin goes down, hunger goes up) and reduced sleep quality, flexibility, and strength. These altered circadian rhythms have also been linked by scientists to premenstrual dysphoric disorder (PMDD), a severe form of premenstrual syndrome that affects up to 8 percent of all premenopausal women. The symptoms include poor-quality sleep, insomnia, depression, tension, intense mood swings, and irritability.

In a Canadian study,[3] researchers compared melatonin secretions and amplitude over a menstrual cycle of a control group and another of PMDD women. They found that the PMDD group's melatonin secretions were lower in all menstrual phases than the control, and that its amplitude was lower during the luteal phase, indicating that women with the disorder have an impaired suprachiasmatic nucleus, the master clock in their brains. For readers who deal with premenstrual circadian fluctuations every month and are annoyed by people who don't believe they have a biological basis, you have my sympathies. Next time someone says that your symptoms are all in your head, say, "You're right. They *are* in my head, right in the suprachiasmatic nucleus."

The chronotype most likely to have luteal phase symptoms like irritability, mood swings, depression, poor sleep, and insomnia:

Hate to say it, but **Dolphins** lose again. It's all about melatonin secretion and amplitude changes that affect the ability to fall asleep and stay asleep. Insomniacs just don't go with the flow, as it were, with any kind of sleep disturbance. I would recommend melatonin supplements during the luteal phase. Take 0.5 to 1.0 mg ninety minutes before lights-out.

THE WINTER RHYTHM

Seasonal affective disorder (SAD), commonly known as the winter blues, is caused by a number of factors, but chiefly diminished exposure to sunlight. Days are shorter, it gets dark earlier, and people are outside less often. When we're inside all the time, soaking in artificial light, our sleep, vitamin D absorption, hormonal (serotonin and melatonin) and metabolic rhythms reset, leading to a host of symptoms[4]:

Mood: anger, anxiety, apathy, general discontent, hopelessness, inability to feel pleasure, loneliness, loss of interest, mood swings, and sadness
Sleep: excess sleepiness, insomnia, and sleep deprivation
Psychological: depression and repeatedly going over thoughts
Whole body: appetite changes, fatigue, restlessness
Behavioral: crying, irritability, and social isolation
Weight: weight gain or weight loss
Cognition: lack of concentration

Ten million Americans go to their doctors for SAD treatment each year. Millions more suffer without seeking care. Who is most likely to fall under the season? In a Polish study[5] of 101 subjects with an average age of twenty-six, researchers created a Winter Blues Scale and asked subjects to rate themselves in season-related fatigue, appetite, energy, sex drive, general malaise, mood, and sociability, among other areas. They correlated the results and compared them to gender and personality traits. As it turns out, women were twice as likely to suffer during the dark months. Subjects who rate high in neuroticism (prone to anxiety and mood swings) and openness (being sensitive and receptive to the new) were significantly more susceptible to SAD as well. The subjects who used an "avoidance-oriented coping style"—meaning they used distraction (for example, abusing drugs and alcohol, overeating, escaping reality with TV or video games)—slept less and reported darker moods and lower energy throughout the season. The study authors called avoidance coping a kind of human hibernation.

Some theorize that too much artificial light in winter causes SAD. Not necessarily. Researchers at the University of Maryland School of Medicine studied[6] a population that doesn't use artificial light in the wintertime at all, Old Order Amish of Lancaster, Pennsylvania. They tested nearly 500 subjects on chronotype and seasonality of mood, and found that morning types were less likely to get winter blues regardless of the lack of electricity. Lions just don't get SAD as often as other types do.

Rhythm within a rhythm: a word on winter weight gain. Packing on pounds will certainly add to depression, and our waistlines do expand during the winter months. According to our evolutionary biology, however, we should actually lose weight in wintertime.

Melatonin is related to appetite, and to the release of "I'm full" hormone leptin and "I'm hungry" hormone ghrelin. Less melatonin release, less feeling full and more hunger. This is one of the main reasons sleep deprivation leads to weight gain.

In spring and summer, melatonin declines, making us less sleepy and more hungry. In the warm months, our primitive ancestors slept a bit less and were hungrier when melatonin went down—a good thing, since food was plentiful. In winter, when melatonin increased, they slept more and ate less—a good thing, since food was scarce. In our modern food-abundant culture, we eat a lot all year round, especially during the winter, when we turn toward serotonin-producing high-carb comfort foods to chase away the blues. Mac and cheese might provide temporary relief from SAD. But it becomes a long-term problem if we gain weight when we should be losing it, and then gain more when our bodies evolve to eat with no restrictions. Wolves are particularly vulnerable to this phenomenon. They tend to indulge in food (and alcohol) to get them through dark moods.

The chronotypes most negatively affected by winter blues:

Don't let the phrase "human hibernation" throw you. Bears are only mildly affected. **Dolphins**, on the other hand, are keenly sensitive to environmental changes and score highest in neuroticism, a trait associated with SAD.

Wolves, the chronotype associated with openness, avoidance coping strategy, and mood disorders like depression even in the warm months, are most likely to suffer from SAD in wintertime. To combat winter blues, act like it's summer. Get outside for as much direct sunlight as possible (especially in the morning), exercise outdoors (you'll warm up quickly), and eat fresh fruit and vegetables. Boost serotonin levels naturally with yoga and meditation.

THE DAYLIGHT SAVING RHYTHM

Honestly, I see no utility in changing the clocks twice a year for daylight saving time (DST). DST is a dinosaur. And, like a dinosaur, it can leave a lot of wreckage behind it.

A German study[7] on the impact of daylight saving time found that, for an eight-week period around the change in our social time (but not solar time), the autumnal time change ("fall back") was easier to take for all chronotypes, but the vernal change ("spring ahead") was a rough shift for all types, especially Wolves.

The spring DST change causes a mini–jet lag effect, like crossing one time zone. The average human needs roughly a day to recover from it. Since DST is now on a weekend schedule, it's easier to take. But, come Monday morning, you still feel a little tired, a little clumsy, and a little irritable. In the days after we spring ahead in March, there's an increase in injuries, traffic accidents[8] and heart attacks.[9]

The chronotypes most negatively affected by daylight saving time:

Dolphins do not do well losing an hour's sleep. They need all the sleep they can get, and sacrificing an hour for sunlight isn't worth the trade.

Bears and Wolves might love to fall back, but springing ahead causes sleep inertia the next day, and the fatigue, irritability, possible accidents, and mistakes that may go with it.

Lions are further isolated by DST—it costs them an additional hour in the morning in the pitch-black wee hours, and they feel tired even earlier in the evening.

To smooth the adjustment to DST, act as if you're flying to a new time zone. Get as much direct sunlight as possible, stick to your on-the-clock eating schedule, even if you're not particularly hungry, and exercise at dusk for the three days after the change.

CHAPTER 16

Chrono-Longevity

Chronorhythms change over time. You might be born with a genetic pre-dilection for waking at dawn or for insomnia, but **the parameters of your bio-time are flexible, depending on your chronological age.**

This book is about adulthood, ages twenty-one to sixty-five. Each of the four groups is relatively diverse. Bears are the majority, at approximately 50 percent of the general population, leaving the remaining 50 percent to be split among Dolphins, Lions, and Wolves. The pie chart for the ages before twenty-one and after sixty-five would be divided quite differently.

Babies behave like Wolves by being most active and alert late into the night and sleeping during the day. There is, however, a more accurate statement to be made about babies and chronotype: Although they are born with an SCN in their brain and with the genetic code to become one of the four types, newborns don't use their biological clocks at all until they are two to three months old.

In utero, babies exist in complete darkness. They have zero exposure to sunlight or artificial light. Their uterine pod and natal lifestyle—pretty much just hanging around in the amniotic fluid—offers them no zeitgebers such as the solar cycle or regular mealtimes. More importantly, a newborn's pineal gland, producer of the bio-time hormones melatonin and serotonin, is not fully developed at birth and doesn't grow to its full size until age two.[1] Until three months old, newborns don't produce melatonin at all.[2] They're operating without a clock.

Babies do get some melatonin in their mother's breast milk.[3] It is soothing and does help them sleep and feel restful, especially at night, when their moms are producing more of it and passing the hormone along. But it's not enough to put them to sleep for a whole night, because babies' bellies empty quickly, and they will need more.

Moms also pass along their stress. In a fascinating study[4] of fifty-two breastfeeding mothers, researchers found a correlation between high concentrations of cortisol in the subjects' milk and "negative affectivity"—fear, sadness, discomfort, anger, frustration, and unsoothability—in female babies. The boy babies were more easygoing. What this has to do with babies and bio-time: When the baby cries and you wake up in the middle of the night with adrenal fight-or-flight hormones flowing, you will pass those on to the baby, and neither of you will be able to fall back to sleep.

Eventually—by three months old—babies' pineal glands are fully mature, and their biological clocks start ticking. Parents who use the power of zeitgebers—morning light exposure, regular mealtimes—will establish a healthy sleep/wake cycle for babies and will help to effect a complete 180—from newborn Wolfishness to Lion-like baby cub behavior.

Toddlers tend to be little Lions, according to a study[5] by researchers at the University of Boulder. They assessed the circadian clock of forty-eight healthy three-year-olds, with the help of the subjects' parents. Sleep diaries were created for each toddler, and then their salivary melatonin was collected to measure their biological sleep-onset time. Nearly 60 percent were rated "definitely a morning type" or "rather a morning type." The rest were in the neither morning nor evening category, with zero subjects rated as "definitely an evening type." Anyone who has parented a toddler knows this already. They jump on your bed at first light, take a nap after lunch, and then crash in mid-evening.

As toddlers get older and grow out of nap time, their chronotype stays Lion-like and then veers toward Bearishness—until adolescence, when it zooms, seemingly overnight, into Wolf territory.

Teenagers are predominantly Wolves. They can sleep until afternoon and stay awake long into the night. Noted chronobiologist Till Roenneberg authored a study[6] that suggested that the shift in bio-time from teenage Wolfishness to adult Bearishness marked the end of adolescence. Roenneberg and his team analyzed the chronotypes and ages of 25,000 German and Swiss subjects by charting their sleep midpoint on free days (weekends and vacations, when you don't have to wake for any particular reason). Wolf majority peaked at age twenty and plummeted by age twenty-five, giving way to a Bear majority, which persisted through adulthood.

Seniors. In Roenneberg's study, the majority orientation changed again after age sixty-five—toward morning orientation (Lion). Seniors are known to have Lion tendencies: waking early, eating dinner early, and feeling tired early. Their attention, executive function, and cognition follow a Lion-like pattern[7] of proficiency in the morning and being more easily distracted and unfocused in the evening. However, if you look at total sleep time and quality of sleep in seniors, a different chronotype pattern emerges. In a study[8] of nearly 1,000 seniors with an average age of seventy-four, researchers compared their subjects' bedtime and total sleep times and discovered that, although morning-oriented seniors went to bed an hour before Bearish seniors, they were actually getting twenty minutes less total sleep. Despite going to bed early to get more sleep, they wind up not getting enough sleep anyway. That sounds awfully Dolphinish to me and fits the sleep patterns of my insomniac patients.

According to the National Institutes of Health, 50 percent of seniors struggle with insomnia. It's common for them to have less Stage 3 and Stage 4 slow wave deep sleep. Seniors naturally produce less human growth hormone and melatonin, resulting in fragmented sleep with multiple wake-ups throughout the night. Secondary causes of insomnia strike the elderly—being on sleep-disruptive medications, having pains that keep them up, immobility, anxiety, and medical problems. All of these factors and conditions contribute to their getting less, lighter, and poor-quality sleep.

We all hope to stay active in our senior years and have high quality of life. One way to increase the likelihood of that is to get on good bio-time for your chronotype *now* to set yourself up for better health in your golden years. If you apply the strategies in this book, you'll get more sleep, lose weight, increase muscle mass, and ward off heart disease and diabetes, conditions that will shorten your life or complicate your old age.

CHRONOTYPE MAJORITY BY AGE

Newborn: Wolf
Toddler: Lion
School Age: Lion/Bear
Teenage: Wolf
Adult: Bear
Senior: Lion/Dolphin

LAST MINUTES

The power of when stems from the latest research in biology and medicine. It's truly the cutting edge of health care.

We know that giving chemotherapy at particular times of day changes the effectiveness of the therapy. Time stamps are starting to appear on blood tests because time of day affects what's in there. The idea of not taking time of day into account with blood samples and medication seems outdated, if not dangerous.

Taking the emerging research a step into the future, I believe that, one day soon, personalized chronorhythms will become just another way to treat and cure diseases.

I'm excited about the data I have presented here, and the opportunity to open your eyes to the power of when. I want you to learn your bio-time rhythm and become healthier and happier. Doing so will affect the health and happiness of those around you. You really can do everything better without having to change what and how you do it.

I am constantly looking for new data and will post regularly on my website, www.thepowerofwhen.com. I want to thank the researchers whose studies I've referenced in this book, as well as those who are laying the groundwork for future study. I'll continue to work with patients in my practice and to study how we can use timing to be stronger, faster, healthier, richer, happier, and more successful, and how we can have better relationships every day.

Begin today. I advise my patients to start with changing the biggies—sleep and wake times, mealtimes, exercise times—and then steadily make new adjustments to their schedule every few days or on a weekly basis.

Or you can dive into a totally new schedule immediately. Check your chronotype's master clock (starting on page 333), and see what you should be doing at this very moment. Is it having coffee? Lucky you! Brew yourself a cup. Is it going to bed? Then put on your pajamas and get in bed. Is it going for a run? Then put on your sneakers and get out there. The schedule for health and happiness is in your hands. Check the time, check your bio-timing, and start using the power of when right this minute!

Master Clocks

Dolphin Master Clock

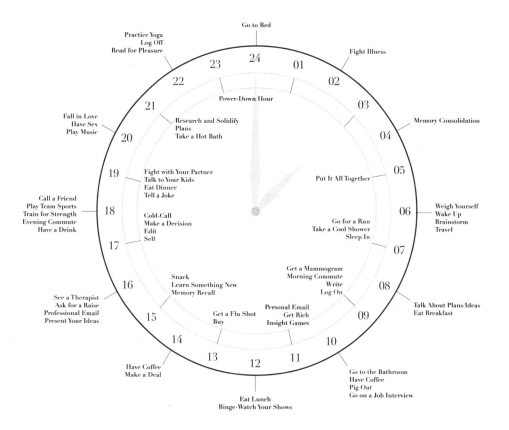

Go to Bed

Practice Yoga
Log Off
Read for Pleasure

Fight Illness

Power-Down Hour

Memory Consolidation

Fall in Love
Have Sex
Play Music

Research and Solidify
Plans
Take a Hot Bath

Put It All Together

Fight with Your Partner
Talk to Your Kids
Eat Dinner
Tell a Joke

Weigh Yourself
Wake Up
Brainstorm
Travel

Call a Friend
Play Team Sports
Train for Strength
Evening Commute
Have a Drink

Cold-Call
Make a Decision
Edit
Sell

Go for a Run
Take a Cool Shower
Sleep In

Get a Mammogram
Morning Commute
Write
Log On

Snack
Learn Something New
Memory Recall

See a Therapist
Ask for a Raise
Professional Email
Present Your Ideas

Get a Flu Shot
Buy

Personal Email
Get Rich
Insight Games

Talk About Plans/Ideas
Eat Breakfast

Have Coffee
Make a Deal

Go to the Bathroom
Have Coffee
Pig Out
Go on a Job Interview

Eat Lunch
Binge-Watch Your Shows

Lion Master Clock

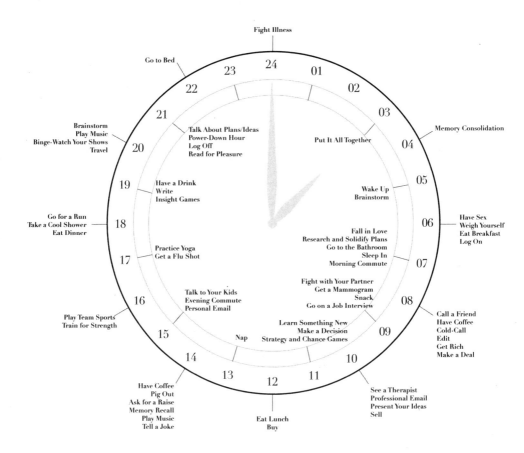

Fight Illness

Go to Bed

24 01
23 02
22 03
21 04

Memory Consolidation

Brainstorm
Play Music
Binge-Watch Your Shows
Travel

20

Talk About Plans/Ideas
Power-Down Hour
Log Off
Read for Pleasure

Put It All Together

19

Have a Drink
Write
Insight Games

05

Wake Up
Brainstorm

Go for a Run
Take a Cool Shower
Eat Dinner

18

06

Have Sex
Weigh Yourself
Eat Breakfast
Log On

17

Practice Yoga
Get a Flu Shot

Fall in Love
Research and Solidify Plans
Go to the Bathroom
Sleep In
Morning Commute

07

Fight with Your Partner
Get a Mammogram
Snack
Go on a Job Interview

16

Talk to Your Kids
Evening Commute
Personal Email

08

Call a Friend
Have Coffee
Cold-Call
Edit
Get Rich
Make a Deal

Play Team Sports
Train for Strength

15

Nap

Learn Something New
Make a Decision
Strategy and Chance Games

09

Have Coffee
Pig Out
Ask for a Raise
Memory Recall
Play Music
Tell a Joke

14

10

See a Therapist
Professional Email
Present Your Ideas
Sell

13 12 11

Eat Lunch
Buy

Bear Master Clock

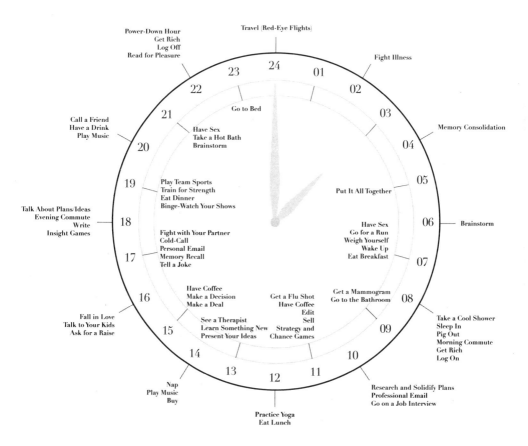

Power-Down Hour
Get Rich
Log Off
Read for Pleasure

Travel (Red-Eye Flights)

Fight Illness

Go to Bed

Call a Friend
Have a Drink
Play Music

Have Sex
Take a Hot Bath
Brainstorm

Memory Consolidation

Play Team Sports
Train for Strength
Eat Dinner
Binge-Watch Your Shows

Put It All Together

Talk About Plans/Ideas
Evening Commute
Write
Insight Games

Brainstorm

Fight with Your Partner
Cold-Call
Personal Email
Memory Recall
Tell a Joke

Have Sex
Go for a Run
Weigh Yourself
Wake Up
Eat Breakfast

Have Coffee
Make a Decision
Make a Deal

Get a Flu Shot
Have Coffee
Edit
Sell
Strategy and
Chance Games

Get a Mammogram
Go to the Bathroom

Fall in Love
Talk to Your Kids
Ask for a Raise

See a Therapist
Learn Something New
Present Your Ideas

Take a Cool Shower
Sleep In
Pig Out
Morning Commute
Get Rich
Log On

Nap
Play Music
Buy

Research and Solidify Plans
Professional Email
Go on a Job Interview

Practice Yoga
Eat Lunch

Wolf Master Clock

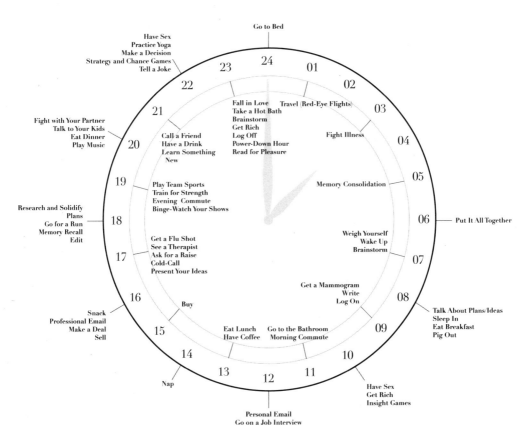

Go to Bed

Have Sex
Practice Yoga
Make a Decision
Strategy and Chance Games
Tell a Joke

23   24   01

22

21

Fall in Love   Travel (Red-Eye Flights)
Take a Hot Bath
Brainstorm
Get Rich
Log Off
Power-Down Hour
Read for Pleasure

02

03

Fight Illness

04

Fight with Your Partner
Talk to Your Kids
Eat Dinner
Play Music

20

Call a Friend
Have a Drink
Learn Something
New

19

Play Team Sports
Train for Strength
Evening Commute
Binge-Watch Your Shows

05

Memory Consolidation

Research and Solidify
Plans
Go for a Run
Memory Recall
Edit

18

Get a Flu Shot
See a Therapist
Ask for a Raise
Cold-Call
Present Your Ideas

06   Put It All Together

17

Weigh Yourself
Wake Up
Brainstorm

07

Get a Mammogram
Write
Log On

08

Snack
Professional Email
Make a Deal
Sell

16

Buy

Talk About Plans/Ideas
Sleep In
Eat Breakfast
Pig Out

15

Eat Lunch   Go to the Bathroom
Have Coffee   Morning Commute

09

14

Nap

13   12   11   10

Have Sex
Get Rich
Insight Games

Personal Email
Go on a Job Interview

Acknowledgments

Valerie Frankel: I can't thank you enough for helping get the thoughts out of my head and onto the page. Your tireless work, research, and help were priceless. But more important to me is our long-standing friendship and the friendship you have with my family. You are AWESOME, and this book is my legacy; thank you for everything.

Mehmet Oz: What can I say? As mentors go, you are at the top of the list. In my darkest hours you were there and showed me the path I could not see. You are an important person in my life in so many ways, words cannot express it, so I will not try. I hope to honor your commitment to me though my accomplishments. Onward, we have a lot of people to help.

Tracy Bear: Your attention to detail, interest, and passion helped me write a better book. Your thoughtful and professional team (Lisa and Zea) have helped this become a reality. Thank you for believing in me and my work; we will change the world.

Alex Glass: Being your first solo client has its advantages. Seriously, you are an amazing individual who "got it" in ways that most others would not. You represent me perfectly and you are the Energizer Bunny of literary agents: you never stop and I love it. Here's to a long future together.

Sandy Climan: Someone once told me that a good agent can get you a good deal, but a great agent can see light where there is darkness. Thanks for seeing the light, and the need for sleep in the world. You are going to help change the world, for the better, in sleep. What an awesome accomplishment. I am grateful for our work together and our friendship.

Craig Cogut: While it is common for a visionary to have great vision, it is uncommon for a visionary to do great things for those around them. You are one of those visionaries who see people and their importance. Thank you for seeing me. My best to Deborah and the boys.

Pegasus Capital: David, Eric, Rick, Alec, and everyone at Pegasus, I just wanted to say a quick thank-you for everything you have done, all of your support, and believing in the power of sleep. This continues to be a great ride, and we *will* change the future one dream at a time.

Arianna Huffington: Thank you for shining a spotlight on sleep. We have worked together for quite a while, and I continue to be amazed by your tireless efforts to be a sleep evangelist. While sleep means so much to you, it means even more to the people whose lives you affect. I hope to continue on the path together and make sleep the next vital sign; it truly is a sleep revolution.

Dave Lakahni: A big thank-you to my new friend and protector. Thank you for watching out for me in the crazy world out there, and this book is just the first step in our future together.

Steven Lockly, PhD: Steven, thank you for continuing to open my eyes to the world of circadian rhythms. Your research is foundational to the science, and without it, this type of book could have never been written. Thank you for all your teachings, your challenges, and beer.

David Cloud: Thank you for all of your support; you and I are in this fight together. I really enjoy our spirited conversations—you are so incredibly knowledgeable, and not just about sleep, but about how to get the world to see sleep for what it truly is: the next vital sign. Bravo, my friend.

Mickey Beyer Clausen: We are two birds of a feather, we love what we do, we know all the pitfalls, and we know how to support each other. You have been an awesome supporter of all my work from day one and always looked out for my best interests. We are aligned spiritually and in friendship. Now if I could only get you to move to LA! Thank you for everything.

Joe Polish: A huge thank-you to you, Joe. You are not only a friend of mine and my family, but a friend to all entrepreneurs out there. Your tireless work, connecting, and curiosity are an inspiration to millions (myself included). Your story is one that many need to hear, and I can't wait to help shout it from the rooftops.

Erin Corbit: Thanks for a great new friendship with someone who feels like an old friend. Your business savvy, bright personality, and awesome energy are contagious.

Six Senses: Neil, Anna, Amber, and the entire Six Senses team, a great big thank-you for all of your inspiration, challenges, and energy in making sleep a luxury for everyone. I am truly honored to be working with the best in the world!

Princess Cruise Lines: Mario, Jason, Danielle, and Trevor, thank you for the opportunity to help millions of passengers get a better night's rest and COME BACK NEW! Thank you for being leaders in wellness and bringing it to the high seas in a spectacular way. I really enjoy working with all of you.

Colin, Mick, Jim, Ben, and ResMed: What can I say about my friends at ResMed? Your company is the leader in every way. The insight, thoughtfulness, and data-driven decisions are the benchmark for everyone to follow. Thank you for believing in me and working so hard together.

Nelly Kim: Nelly, thank you so much for all that you have done to help me see the inside of how business really works. Your talent and sense of humor are equally amazing. I really enjoy working with you in so many ways.

Krystyl Baldwin: What can I say, KB, you simply ROCK! Thanks for being behind the scenes and not running away! Your help has been tremendous throughout this process and so much more. You are incredible.

Grace Tobin: Thank you so much for your awesome illustrations. They help bring out the data in the book beautifully.

Little, Brown: I can't tell you what it means to me that you see my vision and have given me the opportunity to let the world see it as well.

All the amazing researchers that have circadian rhythms as their research interest: Without you this book would never have been written. I am excited to get it to the masses and help them.

Friends at one of the best companies in the world, USANA: We are going to help a lot of people together.

All the journalists who have interviewed me about sleep: Now it is time to talk about wake and circadian rhythms.

All the people who agreed to speak at the first-ever Sleep Success Summit: Sean Croxton; Eric Zalenski, DC; John Bailor; Arianna Huffington; Shawn Stevenson; Terry Carelle, RN; Julie Flygare; Thad Gala, MD; David Cloud; Izabella Wentz, PharmD; Smith Johnston, MD; Jillian Teta; Magdalena Wszelaki; Trevor Cates, MD; Marc Sklar; David Brady; Alan Christiansen, NMD; Alan Greene, MD; Carey Chronis, MD; Michael Murray, NMD; Dan Kalish, NMD; Amy Myers, NMD; Ben Lynch, NMD; Donna Gates; Emily Fletcher; Shiroko Sokitch; Harry Massey; Avocado Wolfe; Josh Axe, DC; Abel James; Dan Pardi; Dave Woynarowski; Russell Friedman; Tom Morter, MD; and Niki Gratix.

Notes

Introduction: Timing Is Everything

1. "Edison's Home Life," *Scientific American*, July 1889. Retrieved from the archives on September 9, 2014.
2. A. Derickson, *Dangerously Sleepy: Overworked Americans and the Cult of Manly Wakefulness* (Philadelphia: University of Pennsylvania Press, 2014), 11.

Chapter 1: What's Your Chronotype?

1. G. L. Ottoni, E. Antoniolli, and D. R. Lara, "The Circadian Energy Scale (CIRENS): Two Simple Questions for a Reliable Chronotype Measurement Based on Energy," *Chronobiology International*, April 2011.
2. Konrad S. Jankowski, "The Role of Temperament in the Relationship Between Morning-Eveningness and Mood," February 2014.
3. Jee In Kang, Chun Il Park, et al., "Circadian Preference and Trait Impulsivity, Sensation-Seeking, and Response Inhibition in Healthy Young Adults," *Chronobiology International*, October 2014.
4. Reka Agnes Haraszti, György Purebl, et al., "Morningness-Eveningness Interferes with Perceived Health, Physical Activity, Diet, and Stress Levels in Working Women," *Chronobiology International*, August 2014.
5. Sirimon Reutrakul, Megan M. Hood, et al., "The Relationship Between Breakfast Skipping, Chronotype, and Glycemic Control in Type 2 Diabetes," *Chronobiology International*, February 2014.
6. Juan Francisco Diaz-Morales and Cristina Escribano, "Circadian Preference and Thinking Style: Implications for School Achievement," *Chronobiology International*, December 2013; Juan Manuel Antúnez, José Francisco Navarro, and Ana Adan, "Circadian Typology and Emotional Intelligence in Healthy Adults," *Chronobiology International*, October 2013.
7. Paolo Maria Russo, Luigi Leone, et al., "Circadian Preference and the Big Five: The Role of Impulsivity and Sensation Seeking," *Chronobiology International*, October 2012.

Chapter 2: A Perfect Day in the Life of a Dolphin

1. Name and identifying details have been changed to protect my patient's privacy.
2. A. Rodenbeck, G. Huether, E. Ruther, and G. Hajak, "Interactions Between Evening

and Nocturnal Cortisol Secretion and Sleep Parameters in Patients with Severe Chronic Primary Insomnia," *Neuroscience Letters*, May 2002.

3. J. Backhaus, K. Junghanns, and F. Hohagen, "Sleep Disturbances Are Correlated with Decreased Morning Awakening Salivary Cortisol," *Psychoneuroendocrinology*, October 2004.

4. S. P. Drummond, M. Walker, E. Almklov, M. Campos, D. E. Anderson, and L. D. Straus, "Neural Correlates of Working Memory Performance in Primary Insomnia," *Sleep*, September 2013. Here and throughout, the affiliation of the lead researcher is noted in the text.

Chapter 3: A Perfect Day in the Life of a Lion

1. Name and identifying details have been changed to protect my patient's privacy.

2. Jessica Rosenberg, Ivan I. Maximov, et al., "'Early to Bed, Early to Rise': Diffusion Tensor Imaging Identifies Chronotype-Specificity," *NeuroImage*, January 2014.

3. Kai-Florian Storch, Ian D. Blum, et al., "A Highly Tunable Dopaminergic Oscillator Generates Ultradian Rhythms of Behavioral Arousal," *eLife*, December 2014.

4. Christina Schmidt, Fabienne Collette, et al., "Homeostatic Sleep Pressure and Responses to Sustained Attention in the Suprachiasmatic Area," *Science*, April 2009.

Chapter 4: A Perfect Day in the Life of a Bear

1. Name and identifying details have been changed to protect my patient's privacy.

Chapter 5: A Perfect Day in the Life of a Wolf

1. Name and identifying details have been changed to protect my patient's privacy.

Chapter 6: Relationships

1. D. Singh and P. M. Bronstad, "Female Body Odour is a Potential Cue to Ovulation," *Proceedings Biological Sciences*, April 2001.

2. Yan Zhang, Fanchang Kong, Yanli Zhong, and Hui Kou, "Personality Manipulations: Do They Modulate Facial Attractiveness Ratings?" *Personality and Individual Differences*, November 2014.

3. Inna Schneiderman, Orna Zagoory-Sharon, James F. Leckman, and Ruth Feldman, "Oxytocin During the Initial Stages of Romantic Attachment: Relations to Couples' Interactive Reciprocity," *Psychoneuroendocrinology*, August 2012.

4. Lisa M. Jaremka, Christopher P. Fagundes, et al., "Loneliness Promotes Inflammation During Acute Stress," *Psychological Science*, July 2013.

5. Melvin C. Washington, Ephraim A. Okoro, and Peter W. Cardon, "Perceptions of Civility for Mobile Phone Use in Formal and Informal Meetings," *Business and Professional Communications Quarterly*, October 2013.

6. T. Aledavood, E. López, et al., "Daily Rhythms in Mobile Telephone Communication." *PLoS ONE*, September 2015.

7. Andrew Steptoe et al., "Social Isolation, Loneliness, and All-Cause Mortality in Older Men and Women," *Proceedings of the National Academy of Science*, February 2013.

8. Eti Ben Simon et al., "Losing Neutrality: The Neural Basis of Impaired Emotional Control without Sleep," *Journal of Neuroscience*, September 2015.

9. B. Baran, E. F. Pace-Schott, C. Ericson, and R. M. Spencer, "Processing of Emotional Reactivity and Emotional Memory over Sleep," *Journal of Neuroscience*, January 2012.

10. Konrad S. Jankowski and W. Ciarkowska, "Diurnal Variation in Energetic Arousal, Tense Arousal, and Hedonic Tone in Extreme Morning and Evening Types," *Chronobiology international*, July 2008.

11. Maciej Stolarski, Maria Ledzińska, and Gerald Matthews, "Morning Is Tomorrow, Evening Is Today: Relationships Between Chronotype and Time Perspective," *Biological Rhythm Review*, February 2012.

12. Roberto Refinetti, "Time for Sex: Nycthemeral Distribution of Human Sexual Behavior," *Journal of Circadian Rhythms*, March 2005.

13. C. Piro, F. Fraioli, P. Sciarra, and C. Conti, "Circadian Rhythm of Plasma Testosterone, Cortisol and Gonadotropins in Normal Male Subjects," *Journal of Steroidal Biochemistry*, May 1973.

14. D. Herbenick et al., "Sexual Behavior in the United States: Results from a National Probability Sample of Men and Women Ages 14–94," *Journal of Sexual Medicine*, October 2010.

15. Konrad S. Jankowski, J. F. Díaz-Morales, and C. Randler, "Chronotype, Gender, and Time for Sex," *Chronobiology International*, October 2014.

16. Mareike B. Wieth and Rose T. Zacks, "Time of Day Effects on Problem Solving: When the Non-optimal Is Optimal," *Thinking & Reasoning,* December, 2011.

17. Answer: I have no idea! I'm a psychologist, not a mathematician.

18. This is an old one that seems hopelessly dated today. The answer is that the surgeon is the boy's mother, of course. The surgeon could also be the boy's other father. But you get the point.

19. Tania Lara, Juan Antonio Madrid, and Ángel Correa, "The Vigilance Decrement in Executive Function Is Attenuated When Individual Chronotypes Perform at Their Optimal Time of Day," *PloS One*, February 2014.

20. Ming-Te Wang and Sarah Kenny, "Longitudinal Links Between Fathers' and Mothers' Harsh Verbal Discipline and Adolescents' Conduct Problems and Depressive Symptoms," *Child Development,* May/June 2014.

21. Cristina Escribano et al., "Morningness/eveningness and School Performance Among Spanish Adolescents: Further Evidence," *Learning and Individual Differences,* June 2012.

22. Y. H. Lin, "Association Between Morningness-Eveningness and the Severity of Compulsive Internet Use: The Moderating Role of Gender and Parenting Style," *Sleep Medicine,* December 2013.

Chapter 7: Fitness

1. Elise Facer-Childs and Roland Brandstaetter, "The Impact of Circadian Phenotype and Time Since Awakening on Diurnal Performance in Athletes," *Current Biology*, February 2015.

2. Scott R. Collier, Kimberly Fairbrother, et al., "Effects of Exercise Timing on Sleep Architecture and Nocturnal Blood Pressure in Prehypertensives," *Vascular Health and Risk Management*, December 2014.

3. F. Guillen and S. Laborde, "Higher-Order Structure of Mental Toughness and the Analysis of Latent Mean Differences Between Athletes from 34 Disciplines and Non-Athletes," *Personality and Individual Differences*, April 2014.

4. J. M. Antúnez, J. F. Navarro, and Ana Adan, "Circadian Typology and Emotional Intelligence in Healthy Adults," *Chronobiology International*, October 2013.

5. A. Shechter and M. P. St-Onge, "Delayed Sleep Timing Is Associated with Low Levels of Free-Living Physical Activity in Normal Sleeping Adults," *Sleep Medicine*, December 2014.

6. Alessandra di Cagno et al., "Time of Day — Effects on Motor Coordination and Reactive Strength in Elite Athletes and Untrained Adolescents," *Journal of Sports Science and Medicine*, March 2013.

7. N. R. Okonta, "Does Yoga Therapy Reduce Blood Pressure in Patients with Hypertension? An Integrative Review," *Holistic Nursing Practice*, May 2012.

8. G. M. Cavallera et al., "Personality, Cognitive Styles, and Morningness-Eveningness Disposition in a Sample of Yoga Trainees," *Medical Science Monitor*, February 2014.

9. J. Reed, "Self-Reported Morningness-Eveningness Related to Positive Affect-Change Associated with a Single Session of Hatha Yoga," *International Journal of Yoga Therapy*, September 2014.

10. M. Sedliak, T. Finni, S. Cheng, M. Lind, and K. Häkkinen, "Effect of Time-of-Day-Specific Strength Training on Muscular Hypertrophy in Men," *Journal of Strength and Conditional Research*, December 2009.

11. L. D. Hayes, G. F. Bickerstaff, and J. S. Baker, "Interactions of Cortisol, Testosterone, and Resistance Training: Influence of Circadian Rhythms," *Chronobiology International*, June 2010.

12. S. P. Bird and K. M. Tarpenning, "Influence of Circadian Time Structure on Acute Hormonal Responses to a Single Bout of Heavy-Resistance Exercise in Weight-Trained Men," *Chronobiology International*, January 2004.

13. Konrad S. Jankowski, "Morning Types are Less Sensitive to Pain Than Evening Types All Day Long," *European Journal of Pain*, August 2013.

Chapter 8: Health

1. A. M. Curtis et al., "Circadian Control of Innate Immunity in Macrophages by miR-155 Targeting Bmal1," *Proceedings of the National Academy of Sciences USA*, May 2015.

2. Mattia Lauriola et al., "Diurnal Suppression of EGFR Signalling by Glucocorticoids and Implications for Tumour Progression and Treatment," *Nature Communications*, October 2014.

3. A. A. Prather, D. Janicki-Deverts, M. H. Hall, and S. Cohen, "Behaviorally Assessed Sleep and Susceptibility to the Common Cold," *Sleep*, 2015.

4. David Gozal et al., "Fragmented Sleep Accelerates Cancer Growth and Progression Through Recruitment of Tumor-Associated Macrophages and TLR4 Signaling," *Cancer Research*, March 2014.

5. J. Aviram, T. Shochat, and D. Pud, "Pain Perception in Healthy Young Men Is Modified by Time-of-Day and Is Modality Dependent," *Pain Medicine*, December 2014.

6. K. M. Edwards and V. E. Burns et al., "Eccentric Exercise as an Adjuvant to Influenza Vaccination in Humans," *Brain, Behavior and Immunity*, February 2007.

7. L. C. Russell, "Caffeine Restriction as Initial Treatment for Breast Pain," *Nurse Practitioner,* February 1989.

8. Diana L. Miglioretti et al., "Accuracy of Screening Mammography Varies by Week of Menstrual Cycle," *Radiology,* February 2011.

9. P. C. Konturek, T. Brzozowski, and S. J. Konturek, "Gut Clock: Implication of Circadian Rhythms in the Gastrointestinal Tract," *Journal of Physiology and Pharmacology,* April 2011.

10. S. R. Brown, P. A. Cann, and N. W. Read, "Effect of Coffee on Distal Colon Function," *Gut,* April 1990.

11. Kok-Ann Gwee, "Disturbed Sleep and Disrupted Bowel Functions: Implications for Constipation in Healthy Individuals," *Journal of Neurogastroenterology and Motility,* April 2011.

12. Francisco Díaz-Morales, Konrad S. Jankowski, Christian Vollmer, and Christoph Randler, "Morningness and Life Satisfaction: Further Evidence from Spain," *Chronobiology International,* October 2013.

13. Bel Bei, Jason C. Ong, Shanthat M. W. Rajaratnam, and Rachel Manber, "Chronotype and Improved Sleep Efficiency Independently Predict Depressive Symptom Reduction After Group Cognitive Behavioral Therapy for Insomnia," *Journal of Clinical Sleep Medicine,* September 2015.

14. Juan Manuel Antúnez, José Francisco Navarro, and Ana Adan, "Circadian Typology and Emotional Intelligence in Healthy Adults," *Chronobiology International,* October 2013.

15. E. D. Buhr, S. H. Yoo, and J. S. Takahashi, "Temperature as a Universal Resetting Cue for Mammalian Circadian Oscillators," *Science,* October 2010.

16. Simone M. Ritter and Ap Dijksterhuis, "Creativity—The Unconscious Foundations of the Incubation Period," *Frontiers in Human Neuroscience,* April 2014.

17. Erhard Haus and Franz Halberg, "24-Hour Rhythm in Susceptibility of C mice to a Toxic Dose of Ethanol," *Journal of Applied Physiology,* November 1959.

18. Tobias Bonten et al., "Effect of Aspirin Intake at Bedtime Versus on Awakening on Circadian Rhythm of Platelet Reactivity: A Randomized Cross-Over Trial," *Thrombosis and Haemostasis Journal,* September 2014.

19. Alan Wallace, David Chinn, and Greg Rubin, "Taking Simvastatin in the Morning Compared with in the Evening: Randomised Controlled Trial," *British Medical Journal,* October 2003.

20. R. C. Hermida, D. E. Ayala, J. R. Fernández, and A. Mojón, "Sleep-Time Blood Pressure: Prognostic Value and Relevance as a Therapeutic Target for Cardiovascular Risk Reduction," *Chronobiology International,* March 2013.

21. Carly R. Pacanowski and David A. Levitsky, "Frequent Self-Weighing and Visual Feedback for Weight Loss in Overweight Adults," *Journal of Obesity,* April 2015.

22. Rena R. Wing, Deborah F. Tate, et al. "A Self-Regulation Program for Maintenance of Weight Loss," *New England Journal of Medicine,* October 2006.

23. E. Culnan, J. D. Kloss, and M. Grandner, "A Prospective Study of Weight Gain Associated with Chronotype Among College Freshmen," *Chronobiology International,* June 2013.

Chapter 9: Sleep

1. Adam T. Wertz et al., "Effects of Sleep Inertia on Cognition," *JAMA*, February 2006.
2. Sara C. Mednick and Sean P. A. Drummond, "Perceptual Deterioration Is Reflected in the Neural Response: fMRI Study Between Nappers and Non-Nappers," *Perception*, June 2009.
3. Amber Brooks and Leon Lack, "A Brief Afternoon Nap Following Nocturnal Sleep Restriction: Which Nap Duration Is Most Recuperative?" *Sleep*, November 2006.
4. S. Asaoka, H. Masaki, K. Ogawa, et al., "Performance Monitoring During Sleep Inertia After a 1-h Daytime Nap," *Journal of Sleep Research*, September 2010.
5. A. Mednick, K. Nakayama, and R. Stickgold, "Sleep-Dependent Learning: A Nap Is as Good as a Night," *Natural Neuroscience*, July 2003.
6. To use the interactive Nap Wheel yourself, go to www.saramednick.com.
7. J. A. Vitale, E. Roveda, et al., "Chronotype Influences Activity Circadian Rhythm and Sleep: Differences in Sleep Quality Between Weekdays and Weekend," *Chronobiology International*, April 2015.
8. T. Roenneberg, K. V. Allebrandt, et al., "Social Jetlag and Obesity," *Current Biology*, May 2012.
9. Michael Parsons et al., "Social Jetlag, Obesity and Metabolic Disorder: Investigation in a Cohort Study," *International Journal of Obesity*, January 2015.
10. Floor M. Kroese, Denise T. D. DeRidder, et al., "Bedtime Procrastination: Introducing a New Area of Procrastination," *Frontiers in Psychology*, June 2014.
11. Jane E. Ferrie, Martin J. Shipley, et al., "A Prospective Study of Change in Sleep Duration: Associations with Mortality in the Whitehall II Cohort," *Sleep*, December 2007.
12. Wendy M. Toxel, "It's More Than Sex: Exploring the Dyadic Nature of Sleep and Implications for Health," *Psychosomatic Medicine*, July 2010.

Chapter 10: Eat and Drink

1. Megumi Hatori, Christopher Vollmers, Satchidananda Panda, et al., "Time-Restricted Feeding without Reducing Caloric Intake Prevents Metabolic Diseases in Mice Fed a High-Fat Diet," *Cell Metabolism*, June 2012.
2. Amandine Chaix, Amir Zarrinpar, Phuong Miu, and Satchidananda Panda, "Time-Restricted Feeding Is a Preventative and Therapeutic Intervention against Diverse Nutritional Challenges," *Cell Metabolism*, December 2014.
3. Storch and Blum, "A Highly Tunable Dopaminergic Oscillator Generates Ultradian Rhythms of Behavioral Arousal."
4. M. Garaulet, P. Gómez-Abellán, et al. "Timing of Food Intake Predicts Weight Loss Effectiveness," *International Journal of Obesity*, January 2013.
5. Leah E. Cahill, Stephanie E. Chiuve, Eric Rimm, et al., "Prospective Study of Breakfast Eating and Incident Coronary Heart Disease in a Cohort of Male US Health Professionals," *Circulation*, May 2013.
6. Tracy L. Rupp, Christine Acebo, and Mary A. Carskadon, "Evening Alcohol Suppresses Salivary Melatonin in Young Adults," *Chronobiology International*, 2007.
7. Christopher B. Forsyth, et al., "Circadian Rhythms, Alcohol and Gut Interactions," *Alcohol*, November 2014.

8. Uduak S. Udoh et al., "The Molecular Circadian Clock and Alcohol-Induced Liver Injury," *Biomolecules*, October 2015.

9. Roger H. L. Wilson, Edith J. Newman, Henry W. Newman, "Diurnal Variation in Rate of Alcohol Metabolism," *Journal of Applied Physiology*, March 1956.

10. C. L. Ruby, A. J. Brager, et al., "Chronic Ethanol Attenuates Circadian Photic Phase Resetting and Alters Nocturnal Activity Patterns in the Hamster," *American Journal of Physiology*, September 2009.

11. G. Prat and A. Adan, "Influence of Circadian Typology on Drug Consumption, Hazardous Alcohol Use, and Hangover Symptoms," *Chronobiology International*, April 2011.

12. William R. Lovallo, Thomas L. Whitsett, et al., "Caffeine Stimulation of Cortisol Secretion Across the Waking Hours in Relation to Caffeine Intake Levels," *Psychosomatic Medicine*, February 2005.

13. Anjalene Whittier, Sixto Sanchez, et al., "Eveningness Chronotype, Daytime Sleepiness, Caffeine Consumption, and Use of Other Stimulants Among Peruvian University Students," *Journal of Caffeine Research*, March 2014.

14. Tina M. Burke, Rachel R. Markwald, et al., "Effects of Caffeine on the Human Circadian Clock in Vivo and in Vitro," *Science Translational Medicine*, September 2015.

15. C. Drake, T. Roehrs, J. Shambroom, and T. Roth, "Caffeine Effects on Sleep Taken 0, 3, or 6 Hours Before Going to Bed," *Journal of Clinical Sleep Medicine*, November 2013.

16. Roy F. Baumeister, Ellen Bratslavsky, Mark Muraven, and Dianne M. Tice, "Ego Depletion: Is the Active Self a Limited Resource?" *Personality Processes and Individual Differences*, May 1998.

17. Frank A. J. L. Scheer, Christopher J. Morris, and Steven A. Shea, "The Internal Circadian Clock Increases Hunger and Appetite in the Evening Independent of Food Intake and Other Behaviors," *Obesity*, March 2013.

18. Laura K. Fonken, Joanna L. Workmann, et al., "Light at Night Increases Body Mass by Shifting the Time of Food Intake," *PNAS*, October 2010.

19. Christopher S. Colwell, Dawn H. Loh, et al., "Misaligned Feeding Impairs Memory," *eLife*, December 2015.

20. Angela Kong, Anne McTiernan, et al., "Associations Between Snacking and Weight Loss and Nutrient Intake Among Postmenopausal Overweight-to-Obese Women in a Dietary Weight Loss Intervention," *Journal of the American Dietary Association*, December 2011.

21. C. A. Crispim, I. Z. Zimberg, B. G. dos Reis, R. M. Diniz, S. Tufik, and M. T. de Mello, "Relationship Between Food Intake and Sleep Pattern in Healthy Individuals," *Journal of Clinical Sleep Medicine,* December 2011.

22. A. Harb, R. Levandovski, et al., "Night Eating Patterns and Chronotypes: A Correlation with Binge Eating Behaviors," *Psychiatry Research* December 2012.

Chapter 11: Work

1. Boris Egloff, Anja Tausch, and Carl-Walter Kohlman, "Relationships Between Time of Day, Day of the Week, and Positive Mood: Exploring the Role of the Mood Measure," *Motivation and Emotion*, January 1995.

2. Sunita Sah, Don A. Moore, and Robert J. MacCoun, "Cheap Talk and Credibility: The

Consequences of Confidence and Accuracy on Advisor Credibility and Persuasiveness," *Organizational Behavior and Human Decision Processes*, July 2013.

3. J. M. Antúnez, J. F. Navarro, A. Adan, "Circadian Topography Is Related to Resilience and Optimism in Healthy Adults," *Chronobiology International*, May 2015.

4. Christoph Randler and Lena Salinger, "Relationship Between Morningness-Eveningness and Temperament and Character Dimensions in Adolescents," *Personality and Individual Differences*, January 2011.

5. Margo Hilbrecht, Bryan Smale, and Steven E. Mock, "Highway to Health? Commute Time and Well-Being Among Canadian Adults," *World Leisure Journal*, April 2014.

6. Ángel Correa, Enrique Molina, and Daniel Sanabria, "Effect of Chronotype and Time of Day on Vigilance Decrement During Simulated Driving," *Accident Analysis and Prevention*, June 2012.

7. Adam Martin, Yevgeniy Goryakin, and Marc Suhrcke, "Does Active Commuting Improve Psychological Wellbeing? Longitudinal Evidence from Eighteen Waves of the British Household Panel Survey," *Preventive Medicine*, December 2014.

8. Juan Francisco Díaz-Morales, Joseph R. Ferrari, and Joseph R. Cohen, "Indecision and Avoidant Procrastination: The Role of Morningness-Eveningness and Time Perspective in Chronic Delay Lifestyles," *Journal of General Psychology*, July 2008.

9. Correa, Lara, and Madrid, "The Vigilance Decrement in Executive Function Is Attenuated When Individual Chronotypes Perform at Their Optimal Time of Day."

10. Farshad Kooti, Luca Maria Aiello, Mihajlo Grbovic, Kristina Lerman, and Amin Mantrach, "Evolution of Conversation in the Age of Email Overload," International World Wide Web Conference Committee, May 2015.

11. M. A. Miller, S. D. Rothenberger, et al., "Chronotype Predicts Positive Affect Rhythms Measured by Ecological Momentary Assessment," *Chronobiology International*, April 2015.

12. R. L. Matchock and J. T. Mordkoff, "Chronotype and Time-of-Day Influences on the Alerting, Orienting, and Executive Components of Attention," *Experimental Brain Research*, January 2009.

13. Uri Simonsohn and Francesca Gino, "Daily Horizons: Evidence of Narrow Bracketing in Judgment From 10 Years of M.B.A. Admissions Interviews," *Psychological Science*, May 2012.

14. L. Carlucci and J. Case, "On the Necessity of U-Shaped Learning," *Topics in Cognitive Science*, January 2013.

15. Pablo Valdez, Candelaria Ramírez, and Aída García, "Circadian Rhythms in Cognitive Performance: Implications for Neuropsychological Assessment," *ChronoPhysiology and Therapy*, December 2012.

16. Christina Schmidt, Fabienne Collette, and Carolin F. Reichert, et al., "Pushing the Limits: Chronotype and Time of Day Modulate Working Memory-Dependent Cerebral Activity," *Frontiers in Neuroscience*, September 2015.

17. Paula Alhola and Päivi Polo-Kantola, "Sleep Deprivation: Impact on Cognitive Performance," *Neuropsychiatric Disease and Treatment*, October 2007.

18. Todd McElroy and David L. Dickinson, "Thoughtful Days and Valenced Nights: How Much Will You Think About the Problem?" *Judgment and Decision Making*, December 2010.

19. M. E. Jewett, J. K. Wyatt, et al. "Time Course of Sleep Inertia Dissipation in Human Performance and Alertness," *Journal of Sleep Research*, March 1999.

20. Díaz-Morales, Ferrari, and Cohen, "Indecision and Avoidant Procrastination: The Role of Morningness-Eveningness and Time Perspective in Chronic Delay Lifestyles."

21. Seung-Schik Yoo, Peter T. Hu, Ninad Gujar, Ferenc A. Jolesz, and Matthew P. Walker, "A Deficit in the Ability to Form New Human Memories Without Sleep," *Nature Neuroscience*, February 2007.

22. Guang Yang, Cora Sau Wan Lai, Joeseph Cichone, Lei Ma, Wei Li, and Wen-Biao Gan, "Sleep Promotes Branch-Specific Formation of Dendritic Spines After Learning," *Science*, June 2014.

23. Alhola and Polo-Kantola. "Sleep Deprivation: Impact on Cognitive Performance."

24. F. F. Barbosa and F. S. Albuquerque, "Effect of the Time-of-Day of Training on Explicit Memory," *Brazilian Journal of Medical and Biological Research*, May 2008.

25. Keith Harris, "A Statistical Analysis of Suggested and Accepted Times for Meetings and Events," WhenIsGood.net, October 2009.

26. Paul King and Ralph Behnke, "Patterns of State Anxiety in Listening Performance," *Southern Communication Journal*, 2004.

Chapter 12: Creativity

1. B. J. Shannon, R. A. Dosenbach, et al., "Morning-Evening Variation in Human Brain Mechanism and Memory Circuits," *Journal of Neurophysiology*, March 2013.

2. Wieth and Zacks, "Time of Day Effects on Problem Solving: When the Non-Optimal Is Optimal."

3. Floris T. Van Vugt, Katharina Treutler, Eckart Altenmüller, and Hans-Christian Jabusch, "The Influence of Chronotype on Making Music: Circadian Fluctuations in Pianists' Fine Motor Skills," *Frontiers in Human Neuroscience*, July 2013.

4. Ruth Wells, Tim Outhred, James A. J. Heathers, Daniel S. Quintana, Andrew H. Kemp, "Matter Over Mind: A Randomised-Controlled Trial of Single-Session Biofeedback Training on Performance Anxiety and Heart Rate Variability in Musicians," *PLOS One*, October 2012.

5. Jeffrey M. Ellenbogen, Peter T. Hu, Jessica D. Payne, Debra Titone, and Matthew P. Walker, "Human Relational Memory Requires Time and Sleep," *PNAS*, March 2007.

6. For a list of hundreds of RAT questions and to self-test, go to www.remote-associates-test.com.

7. cheese

8. ice

9. sore

10. brush

11. camp

12. paper

13. pack

14. play

15. gate

16. Denise J. Cai, Sarnoff A. Mednick, Elizabeth M. Harrison, Jennifer C. Kanady, and Sara C. Mednick, "REM, not Incubation, Improves Creativity by Priming Associative Networks," *PNAS*, May 2009.

17. Wieth and Zacks, "Time of Day Effects on Problem Solving: When the Non-Optimal Is Optimal."
18. He split the rope lengthwise and tied the ends together.
19. In ancient times, they didn't use "BC."
20. Day fifty-nine. Since the flowers double every day, it'd be half covered the day before it was fully covered on day sixty.
21. Bob is twelve; his father, thirty-six. Four years ago, Bob was eight and his father was thirty-two.
22. Monday: Bill, tacos. Tuesday: Dave, steak. Wednesday: Carl, pizza. Thursday: Eric, fish. Friday: Andy, Thai.
23. Anna, Charlie, carnation. Isabel, Tom, daffodil. Yvonne, Ken, lily. Emily, Ron, rose.
24. Andrew F. Jarosz, Gregory J. H. Colflesh, and Jennifer Wiley, "Uncorking the Muse: Alcohol Intoxication Facilitates Creative Problem Solving," *Consciousness and Cognition*, March 2012.
25. Most of this information is from Mason Currey's fascinating book *Daily Rituals: How Artists Work* (New York: Knopf, 2013).
26. A link to the video here: http://juliacameronlive.com/basic-tools/morning-pages/.

Chapter 13: Money

1. Karen J. Pine and Ben Fletcher, "Women's Spending Behavior Is Menstrual-Cycle Sensitive," *Personality and Individual Differences*, January 2011.
2. Oliver B. Büttner, Anna Marie Shultz, et al., "Hard to Ignore: Impulsive Buyers Show an Attentional Bias in Shopping Situations," *Social Psychological & Personality Science*, April 2014.
3. Benjamin G. Serfas, Oliver B. Büttner, and Arnd Florack, "Eyes Wide Shopped: Shopping Situations Trigger Arousal in Impulsive Buyers," *PLOS One*, December 2014.
4. Janowski and Ciarkowska, "Diurnal Variation in Energetic Arousal, Tense Arousal, and Hedonic Tone in Extreme Morning and Evening Types."
5. Scott I. Rick, Beatriz Pereira, and Katherine A. Burson, "The Benefits of Retail Therapy: Making Purchase Decisions Reduces Residual Sadness," *Journal of Consumer Psychology*, July 2014.
6. Shaun A. Saunders, Michael W. Allen, and Kay Pozzebon, "An Exploratory Look at the Relationship Between Materialistic Values and Goals and Type A Behaviour," *Journal of Pacific Rim Psychology*, June 2008.
7. Ayalla Ruvio, Eli Somer, and Aric Rindfleisch, "When Bad Gets Worse: The Amplifying Effect of Materialism on Traumatic Stress and Maladaptive Consumption," *Journal of the Academy of Marketing Science*, January 2014.
8. Alison Jing Xu, Norbert Schwarz, and Robert S. Wyer Jr., "Hunger Promotes Acquisition of Nonfood Objects," *Proceedings of the National Academy of Sciences*, January 2015.
9. John Kounios and Mark Beeman, "The Aha! Moment: The Cognitive Neuroscience of Insight," *Current Directions in Psychological Science*, August 2009.
10. Mary Helen Immordino-Yang, Joanna A. Christodoulous, and Vanessa Singh, "Rest Is Not Idleness: Implications of the Brain's Default Mode for Human Development and Education," *Perspectives on Psychological Science*, July 2012.

11. Davide Ponzi, M. Claire Wilson, and Dario Maestripieri, "Eveningness is Associated with Higher Risk-Taking, Independent of Sex and Personality," *Psychological Reports: Sociocultural Issues in Psychology,* December 2014.

12. Brian Gunia, Christopher M. Barnes, and Sunita Sah, "Larks and Owls: Unethical Behavior Depends on Chronotype as Well as Time-of-Day," *Psychological Science,* December 2014.

13. Shai Danziger, Jonathan Levav, and Liora Avnaim-Pesso, "Extraneous Factors in Judicial Decisions," *PNAS,* February 2011.

14. Carmel Sofer, Ron Dotsch, Daniel H. J. Wigboldus, and Alexander Todorov, "What Is Typical Is Good: The Influence of Face Typicality on Perceived Trustworthiness," *Psychological Science,* December 2014.

15. Tina Sundelin, Mats Lekander, et al., "Cues of Fatigue: Effects of Sleep Deprivation on Facial Appearance," *Sleep,* September 2013.

16. Peter A. Bos, Erno J. Hermans, Nick F. Ramsey, and Jack van Honk, "The Neural Mechanisms by Which Testosterone Acts on Interpersonal Trust," *NeuroImage,* July 2012.

17. Taiki Takahashi, Koki Ikeda, et al. "Interpersonal Trust and Social Stress-induced Cortisol Elevation," *NeuroReport,* February 2005.

18. Aída García, Candelaria Ramirez, Benito Martinez, and Pablo Valdez, "Circadian Rhythms in Two Components of Executive Functions: Cognitive Inhibition and Flexibility," *Biological Rhythm Research,* January 2012.

19. Hichem Slama, Gaétane Deliens, Rémy Schmitz, et al., "Afternoon Nap and Bright Light Exposure Improve Cognitive Flexibility Post Lunch," *PLoS One,* May 2015.

Chapter 14: Fun

1. Yoon Hi Sung, Eun Yeon Kang, and Wei-Na Lee, "A Bad Habit for Your Health? An Exploration of Psychological Factors for Binge-Watching Behavior," *All Academia Inc,* January 2015.

2. I. N. Fossum, L. T. Nordnes, S. S. Storemark, et al., "The Association Between Use of Electronic Media in Bed Before Going to Sleep and Insomnia Symptoms, Daytime Sleepiness, Morningness, and Chronotype," *Behaviors in Sleep Medicine,* September 2014.

3. Jacob M. Burmeister and Robert A Carels, "Television Use and Binge Eating in Adults Seeking Weight Loss Treatment," *Eating Behaviors,* January 2014.

4. Take the full version at http://personality-testing.info/tests/SD3/.

5. Peter K. Jonason, Amy Jones, and Minna Lyons, "Creatures of the Night: Chronotypes and the Dark Triad Traits," *Personality and Individual Differences,* September 2013.

6. Agata Blachnio, Aneta Przepiorka, and Juan F. Díaz-Morales, "Facebook Use and Chronotype: Results of a Cross-Sectional Study," *Chronobiology International,* August 2015.

7. Masahiro Toda, Nobuhiro Nishio, Satoko Ezoe, and Tatsuya Takeshita, "Chronotype and Smartphone Use Among Japanese Medical Students," *International Journal of Cyber Behavior,* December 2015.

8. L. J. Hadlington, "Cognitive Failures in Daily Life: Exploring the Link with Internet Addiction and Problematic Mobile Phone Use," *Computers in Human Behavior,* October 2015.

9. Fossum, Nordnes, Storemark, et al., "The Association Between Use of Electronic Media in Bed Before Going to Sleep and Insomnia Symptoms, Daytime Sleepiness, Morningness, and Chronotype."

10. Y. H. Lin and S. S. Gau, "Association Between Morningness-Eveningness and the Severity of Compulsive Internet Use: The Moderating Role of Gender and Parenting Style," *Sleep Medicine*, December 2013.

11. Baumeister, Bratslavsky, Muraven, and Tice, "Ego Depletion: Is the Active Self a Limited Resource?"

12. Scott A. Golder and Michael W. Macy. "Diurnal and Seasonal Mood Vary with Work, Sleep, and Daylength Across Diverse Cultures," *Science*, September 2011.

13. Nemanja Spasojevic, Zhisheng Li, Adithya Rao, and Prantik Bhattacharyya. "When to Post on Social Networks," Lithium Technolgies / Klout, June 2015.

14. Gunia, Barnes, and Sah, "Larks and Owls: Unethical Behavior Depends on Chronotype as Well as Time-of-Day."

15. Stephanie D. Womack, Joshua N. Hook, Samuel H Reyna, and Marciana Ramos, "Sleep Loss and Risk-Taking Behavior: A Review of the Literature," *Behavioral Sleep Medicine*, January 2013.

16. W. D. S. Killgore, T. J. Balkin, and N. J. Wesensten, "Impaired Decision Making Following 49 h of Sleep Deprivation," *Journal of Sleep Research*, February 2006.

17. Lili Wang and Tanya L. Chartrand, "Morningness-Eveningness and Risk Taking," *Journal of Psychology*, April 2014.

18. Christian Vollmer, Christoph Randler, Mehmet Barış Horzum, and Tuncay Ayas, "Computer Game Addiction in Adolescents and Its Relationship to Chronotype and Personality," *Sage Open*, January 2014.

19. Peter J. McCormick et al., "Circadian-Related Heteromerization of Adrenergic and Dopamine D_4 Receptors Modulates Melatonin Synthesis and Release in the Pineal Gland," *PLOS Biology*, June 2012.

20. David Comer Kidd and Emanuele Castano, "Reading Literary Fiction Improves Theory of Mind," *Science*, October 2013.

21. Robert S. Wilson, Patricia A. Boyle, Lei Yu, Lisa L. Barneş, Julie A. Schneider, and David A. Bennett, "Life-Span Cognitive Activity, Neuropathologic Burden, and Cognitive Aging," *Neurology*, July 2013.

22. Raymond A. Mar, Keith Oatley, and Jordan B. Peterson, "Exploring the Link Between Reading Fiction and Empathy: Ruling out Individual Differences and Examining Outcomes," *Communications*, January 2009.

23. Anne-Maria Chang, Daniel Aeschbach, Jeanne F. Duffy, and Charles A. Czeisler, "Evening Use of Light-Emitting eReaders Negatively Affects Sleep, Circadian Timing, and Next-Morning Alertness," *PNAS*, January 2015.

24. W. D. Killgore, S. A. McBride, D. B. Killgore, and T. J. Balkin, "The Effects of Caffeine, Dextroamphetamine, and Modafinil on Humor Appreciation During Sleep Deprivation," *Sleep*, June 2006.

25. C. Randler, "Evening Types Among German University Students Score Higher on Sense of Humor After Controlling for Big Five Personality Factors," *Psychological Reports*, October 2008.

26. With your doctor's approval, of course.

27. Specifically, NASA's Multilateral *Guidelines for Management of Circadian Desynchrony in ISS Operations SSP 50480-ANX3*, composed by the Multilateral Medical Operations Panel Spaceflight Human Behavior and Performance Working Group Fatigue Management Team.
28. With your doctor's approval.
29. With your doctor's approval.

Chapter 15: Chrono-Seasonality

1. Christian Cajochen, Songül Altanay-Ekici, Mirjam Münch, Sylvia Frey, Vera Knoblauch, and Anna Wirz-Justice, "Evidence that the Lunar Cycle Influences Human Sleep," *Current Biology*, August 2013.
2. F. C. Baker and H. S. Driver, "Circadian Rhythms, Sleep, and the Menstrual Cycle," *Sleep Medicine*, September 2007.
3. Ari Shechter, Paul Lespérance, N. M. K. Ng Ying Kin, and Diane B. Boivin, "Pilot Investigation of the Circadian Plasma Melatonin Rhythm Across the Menstrual Cycle in a Small Group of Women with Premenstrual Dysphoric Disorder," *PLoS One*, December 2012.
4. Symptom list per the Mayo Clinic.
5. H. Oginska and K. Oginska-Bruchal, "Chronotype and Personality Factors of Predisposition to Seasonal Affective Disorder," *Chronobiology International*, May 2014.
6. Layan Zhang, Daniel S. Evans, et al., "Chronotype and Seasonality: Morningness Is Associated with Lower Seasonal Mood and Behavior Changes in the Old Order Amish," *Affective Disorders*, March 2015.
7. Thomas Kantermann, Myriam Juda, Martha Merrow, and Till Roenneberg, "The Human Circadian Clock's Seasonal Adjustment Is Disrupted by Daylight Saving Time," *Current Biology*, November 2007.
8. Jason Varughese and Richard P Allen, "Fatal Accidents Following Changes in Daylight Savings Time: The American Experience," *Sleep Medicine*, January 2001.
9. Imre Janszky and Rickard Ljung, "Shifts to and from Daylight Saving Time and Incidence of Myocardial Infarction," *New England Journal of Medicine*, October 2008.

Chapter 16: Chrono-Longevity

1. Masayuki Sumida, A. James Barkovich, and T. Hans Newton, "Development of the Pineal Gland: Measurement with MR," *American Journal of Neuroradiology*, February 1996.
2. D. J. Kennaway, G. E. Stamp, F. C. Goble. "Development of Melatonin Production in Infants and the Impact of Prematurity," *The Journal of Clinical Endocrinology and Metabolism*, July 2013.
3. A. Cohen Engler, A. Hadash, N. Shehadeh, and G. Pillar, "Breastfeeding May Improve Nocturnal Sleep and Reduce Infantile Colic: Potential Role of Breast Milk Melatonin," *European Journal of Pediatrics*, April 2012.
4. K. R. Grey, E. P. Davis, C. A. Sandman, and L. M. Glynn, "Human Milk Cortisol Is Associated with Infant Temperament," *Psychoneuroendocrinology*, July 2013.

5. C. T. Simpkin, O. G. Jenni, M. A. Carskadon, K. P. Wright Jr., L. D. Akacem, K. G. Garlo, and M. K. LeBourgeois, "Chronotype Is Associated with the Timing of the Circadian Clock and Sleep in Toddlers," *Journal of Sleep Research*, August 2014.

6. Till Roenneberg, Tim Kuehnle, Peter P. Pramstaller, Jan Ricken, Miriam Havel, Angelika Guth, and Martha Merrow, "A Marker for the End of Adolescence," *Current Biology*, December 2004.

7. J. A. Anderson, K. L. Campbell, T. Amer, C. L. Grady, and L. Hasher, "Timing Is Everything: Age Differences in the Cognitive Control Network Are Modulated by Time of Day," *Psychology and Aging*, September 2014.

8. Timothy H. Monk and Daniel J. Buysse, "Chronotype, Bed Timing, and Total Sleep Time in Seniors," *Chronobiology International*, June 2014.

Index

Note: Italic page numbers refer to illustrations.

About the Author

Michael J. Breus, PhD, is a clinical psychologist and both a Diplomate of the American Board of Sleep Medicine and a Fellow of the American Academy of Sleep Medicine. He was one of the youngest people to have passed the Sleep Medicine Board, at age thirty-one, and, with a specialty in sleep disorders, is one of only 163 psychologists in the world with his credentials and distinction. Dr. Breus is on the clinical advisory board of *The Dr. Oz Show* and is a regular contributor on the show.

As the subject of sleep continues to gain momentum in our sleep-deprived society, Dr. Breus has become a widely recognized leader in this ever-evolving field. He has supplied his expertise as a consultant and a sleep educator (spokesperson) to brands such as Advil PM, Breathe Rite, Crowne Plaza Hotels, DONG Energy (Denmark), Merck (Belsomra), and many more. Among his numerous national media appearances, Dr. Breus has been interviewed on CNN, *The Oprah Winfrey Show, The View, Anderson Cooper 360°, The Rachael Ray Show, Fox & Friends, The Doctors, The Joy Behar Show, CBS This Morning,* the *Today* show, and *Live with Kelly and Michael.* He is an expert resource for most major publications, doing more than a hundred interviews per year (*Wall Street Journal, New York Times, Washington Post,* and most popular magazines). He also appears regularly on *Dr. Oz* and Sirius XM Radio, and has served as the sleep expert for WebMD for over fourteen years.

Dr. Breus has lectured hundreds of times for Fortune 500 and 100 companies, to executive teams, and to entire employee populations. He works with employee health groups to help ensure that sleep health is a priority, offering scalable solutions, and has provided editorial services for numerous medical and psychology peer-reviewed journals.

He is the author of *The Sleep Doctor's Diet Plan* and *Beauty Sleep.*

Dr. Breus has been in private practice for sixteen years and currently practices sleep medicine in Los Angeles. He lives with his wife, son, daughter, two dogs, and a cat in Manhattan Beach, California.